T0259459

WELL LOGGING AND FORMATION EVALUATION

WELL LOGGING AND FORMATION EVALUATION

TOBY DARLING

ELSEVIER

AMSTERDAM • BOSTON • HEIDELBERG • LONDON
NEW YORK • OXFORD • PARIS • SAN DIEGO
SAN FRANCISCO • SINGAPORE • SYDNEY • TOKYO

Gulf Professional Publishing is an imprint of Elsevier

G|P
P|𝖶

Gulf Professional Publishing is an imprint of Elsevier
30 Corporate Drive, Suite 400, Burlington, MA 01803, USA
Linacre House, Jordan Hill, Oxford OX2 8DP, UK

Library of Congress Cataloging-in-Publication Data
Application submitted.

British Library Cataloguing-in-Publication Data
A catalogue record for this book is available from the British Library..

ISBN-13: 978-0-7506-7883-4
ISBN-10: 0-7506-7883-6

For information on all Gulf Professional Publishing
publications visit our Web site at www.books.elsevier.com

Working together to grow libraries in developing countries

www.elsevier.com | www.bookaid.org | www.sabre.org

ELSEVIER   BOOK AID
             International   Sabre Foundation

Transferred to Digital Printing 2012

CONTENTS

INTRODUCTION

The purpose of this book is to provide a series of techniques which will be of real practical value to petrophysicists in their day-to-day jobs. These are based on my experience from many years working in oil companies. To this end I have concentrated wherever possible on providing one recommended technique, rather than offer the reader a choice of different options.

The primary functions of a petrophysicist are to ensure that the right operational decisions are made during the course of drilling and testing a well—from data gathering, completion and testing—and thereafter to provide the necessary parameters to enable an accurate static and dynamic model of the reservoir to be constructed. Lying somewhere between Operations, Production Geology, Seismology, Production Technology and Reservoir Engineering, the petrophysicist has a key role in ensuring the success of a well, and the characterization of a reservoir.

The target audience for this book are operational petrophysicists in their first few years within the discipline. It is expected that they have some knowledge of petroleum engineering and basic petrophysics, but lack experience in operational petrophysics and advanced logging techniques. The book also may be useful for those in sister disciplines (particularly production geology and reservoir engineering) who are using the interpretations supplied by petrophysicists.

C H A P T E R   1

BASICS

1.1 TERMINOLOGY

Like most professions, petroleum engineering is beset with jargon. Therefore, it will make things simpler if I first go through some of the basic terms that will be used throughout this book. Petroleum engineering is principally concerned with building static and dynamic models of oil and gas reservoirs.

Static models are concerned with characterizing and quantifying the structure prior to any production from the field. Hence, key parameters that the models aim to determine are:

- STOIIP = stock tank oil initially in place; usually measured in stock tank barrels (stb)
- GIIP = gas initially in place; usually measured in billion standard cubic feet (Bcf)
- GBV = gross bulk volume; the total rock volume of the reservoir containing hydrocarbon
- NPV = net pore volume; the porespace of the reservoir
- HCPV = hydrocarbon pore volume; the porespace actually containing hydrocarbon
- ϕ = porosity; the proportion of the formation that contains fluids
- k = permeability; usually expressed in millidarcies (md)
- S_w = water saturation; the proportion of the porosity that contains water
- S_h = hydrocarbon saturation; the proportion of the porosity that contains hydrocarbon
- FWL = free water level; the depth at which the capillary pressure in the reservoir is zero; effectively the depth below which no producible hydrocarbons will be found

- HWC = hydrocarbon/water contact; the depth below which the formation is water bearing as encountered in a particular well. Likewise, OWC for oil and GWC for gas
- GOC = gas oil contact; the depth below which any gas in the reservoir will be dissolved in the oil
- Gross thickness = the total thickness of the formation as encountered in a particular well
- Net thickness = the part of the gross thickness that contains porous rock subject to given cutoff criteria
- Pay thickness = the part of the net thickness that is considered to be capable of producing hydrocarbons in a particular well

Because of inherent uncertainties in all the parameters used to determine STOIIP or GIIP, geologists will usually develop probabilistic models, in which all the parameters are allowed to vary according to distribution functions between low, expected, and high values. The resulting static models may then be analyzed statistically to generate the following values, which are used for subsequent economic analyses:

- P_{50} STOIIP: the value of the STOIIP for which there is a 50% chance that the true value lies either above or below the value
- P_{15} STOIIP: the value of the STOIIP for which there is only a 15% chance that the true value exceeds the value. Often called the *high case*.
- P_{85} STOIIP: the value of the STOIIP for which there is an 85% chance that the true value exceeds the value. Often called the *low case*.
- Expected STOIIP: the value of the STOIIP derived by taking the integral of the probability density function for the STOIIP times the STOIIP. For a symmetric distribution, this will equal the P_{50} value.

Similar terminology applies to GIIP.

In order to predict the hydrocarbons that may be actually produced from a field (the reserves), it is necessary to construct a dynamic model of the field. This will generate production profiles for individual wells, subject to various production scenarios. Additional terminology that comes into play includes:

- Reserves = the part of the STOIIP or GIIP that may be actually produced for a given development scenario. Oil companies have their own rules for how reserves are categorized depending on the extent to which they are regarded as proven and accessible through wells. Terms frequently used are *proven reserves, developed reserves, scope for recovery reserves, probable reserves,* and *possible reserves.*

- Remaining reserves = that part of the reserves that has not yet been produced
- Cumulative production = that part of the reserves that has already been produced
- UR = ultimate recovery; the total volume of reserves that will be produced prior to abandonment of the field
- NPV = net present value; the future economic value of the field, taking into account all future present value costs and revenues
- RF = recovery factor; the reserves as a proportion of the STOIIP (or GIIP)
- B_o = oil volume factor; the factor used to convert reservoir volumes of oil to surface (stock tank) conditions. Likewise B_g for gas.

In order to produce the hydrocarbons, wells are needed and a development strategy needs to be constructed. This strategy will typically be presented in a document called the **field development plan** (FDP), which contains a summary of current knowledge about the field and the plans for future development.

Once an FDP has been approved, the drilling campaign will consist of well proposals, in which the costs, well trajectory, geological prognosis, and data-gathering requirements are specified. The petrophysicist plays a part in the preparation of the well proposal in specifying which logs need to be acquired in the various hole sections.

1.2 BASIC LOG TYPES

Below is a list of the main types of logs that may be run, and why they are run.

1.2.1 Logging While Drilling (LWD)

Traditionally, petrophysicists were concerned only with wireline logging, that is, the data acquired by running tools on a cable from a winch after the hole had been drilled. However, advances in drilling/logging technology have allowed the acquisition of log data via tools placed in the actual drilling assembly. These tools may transmit data to the surface on a real-time basis or store the data in a downhole memory from which it may be downloaded when the assembly is brought back to the surface.

LWD tools present a complication for drilling, as well as additional expense. However, their use may be justified when:

- Real-time information is required for operational reasons, such as steering a well (e.g., a horizontal trajectory) in a particular formation or picking of formation tops, coring points, and/or casing setting depths
- Acquiring data prior to the hole washing out or invasion occurring
- Safeguarding information if there is a risk of losing the hole
- The trajectory is such as to make wireline acquisition difficult (e.g., in horizontal wells)

LWD data may be stored downhole in the tools memory and retrieved when the tool is brought to the surface and/or transmitted as pulses in the mud column in real time while drilling. In a typical operation, both modes will be used, with the memory data superseding the pulsed data once the tool is retrieved. However, factors that might limit the ability to fully use both sets of data are:

- Drilling mode: Data may be pulsed only if the drillstring is having mud pumped through it.
- Battery life: Depending on the tools in the string, tools may work in memory mode only between 40 and 90 hours.
- Memory size: Most LWD tools have a memory size limited to a few megabytes. Once the memory is full, the data will start to be overwritten. Depending on how many parameters are being recorded, the memory may become full within 20–120 hours.
- Tool failure: It is not uncommon for a fault to develop in the tool such that the pulse data and/or memory data are not transmissible/recordable.

Some of the data recorded may be usable only if the toolstring is rotating while drilling, which may not always be the case if a steerable mud motor is being used. In these situations, the petrophysicist may need to request drilling to reacquire data over particular intervals while in reaming/rotating mode. This may also be required if the rate of penetration (ROP) has been so high as to affect the accuracy of statistically based tools (e.g., density/neutron) or the sampling interval for tools working on a fixed time sampling increment.

Another important consideration with LWD tools is how close to the bit they may be placed in the drilling string. While the petrophysicist will obviously want the tools as close to the bit as possible, there may be limitations placed by drilling, whose ability to steer the well and achieve a high ROP is influenced by the placement of the LWD toolstring. LWD data that may typically be acquired include the following:

- GR: natural gamma ray emission from the formation
- Density: formation density as measured by gamma ray Compton scattering via a radioactive source and gamma ray detectors. This may also include a photoelectric effect (Pe) measurement.
- Neutron porosity: formation porosity derived from the hydrogen index (HI) as measured by the gamma rays emitted when injected thermal or epithermal neutrons from a source in the string are captured in the formation
- Sonic: the transit time of compressional sound waves in the formation
- Resistivity: the formation resistivity for multiple depths of investigation as measured by an induction-type wave resistivity tool

Some contractors offer LWD-GR, -density, and -neutron as separate up/down or left/right curves, separating the contributions from different quadrants in the borehole. These data may be extremely useful in steering horizontal wells, where it is important to determine the proximity of neighboring formation boundaries before they are actually penetrated. Resistivity data may also be processed to produce a borehole resistivity image, useful for establishing the stratigraphic or sedimentary dip and/or presence of fractures/vugs.

Other types of tool that are currently in development for LWD mode include nuclear magnetic resonance (NMR), formation pressure, and shear sonic.

1.2.2 Wireline Openhole Logging

Once a section of hole has been completed, the bit is pulled out of the hole and there is an opportunity to acquire further openhole logs either via wireline or on the drillstring before the hole is either cased or abandoned. Wireline versions of the LWD tools described above are available, and the following additional tools may be run:

- Gamma ray: This tool measures the strength of the natural radioactivity present in the formation. It is particularly useful in distinguishing sands from shales in siliciclastic environments.
- Natural gamma ray spectroscopy: This tool works on the same principal as the gamma ray, although it separates the gamma ray counts into

three energy windows to determine the relative contributions arising from (1) uranium, (2) potassium, and (3) thorium in the formation. As described later in the book, these data may be used to determine the relative proportions of certain minerals in the formation.

- Spontaneous potential (SP): This tool measures the potential difference naturally occurring when mud filtrate of a certain salinity invades the formation containing water of a different salinity. It may be used to estimate the extent of invasion and in some cases the formation water salinity.
- Caliper: This tool measures the geometry of the hole using either two or four arms. It returns the diameter seen by the tool over either the major or both the major and minor axes.
- Density: The wireline version of this tool will typically have a much stronger source than its LWD counterpart and also include a Pe curve, useful in complex lithology evaluation.
- Neutron porosity: The "standard" neutron most commonly run is a thermal neutron device. However, newer-generation devices often use epithermal neutrons (having the advantage of less salinity dependence) and rely on minitron-type neutron generators rather than chemical sources.
- Full-waveform sonic: In addition to the basic compressional velocity (V_p) of the formation, advanced tools may measure the shear velocity, Stonely velocity, and various other sound modes in the borehole, borehole/formation interface, and formation.
- Resistivity: These tools fall into two main categories: laterolog and induction type. Laterolog tools use low-frequency currents (hence requiring water-based mud [WBM]) to measure the potential caused by a current source over an array of detectors. Induction-type tools use primary coils to induce eddy currents in the formation and then a secondary array of coils to measure the magnetic fields caused by these currents. Since they operate at high frequencies, they can be used in oil-based mud (OBM) systems. Tools are designed to see a range of depths of investigation into the formation. The shallower readings have a better vertical resolution than the deep readings.
- Microresistivity: These tools are designed to measure the formation resistivity in the invaded zone close to the borehole wall. They operate using low-frequency current, so are not suitable for OBM. They are used to estimate the invaded-zone saturation and to pick up bedding features too small to be resolved by the deeper reading tools.
- Imaging tools: These work either on an acoustic or a resistivity principle and are designed to provide an image of the borehole wall that may

be used for establishing the stratigraphic or sedimentary dip and/or presence of fractures/vugs.

- Formation pressure/sampling: Unlike the above tools, which all "log" an interval of the formation, formation-testing tools are designed to measure the formation pressure and/or acquire formation samples at a discrete point in the formation. When in probe mode, such tools press a probe through the mudcake and into the wall of the formation. By opening chambers in the tool and analyzing the fluids and pressures while the chambers are filled, it is possible to determine the true pressure of the formation (as distinct from the mud pressure). If only pressures are required (pretest mode), the chambers are small and the samples are not retained. For formation sampling, larger chambers are used (typically $2^3/_4$ or 6 gallons), and the chambers are sealed for analysis at the surface. For some tools, a packer arrangement is used to enable testing of a discrete interval of the formation (as opposed to a probe measurement), and various additional modules are available to make measurements of the fluid being sampled downhole.

- Sidewall sampling: This is an explosive-type device that shoots a sampling bullet into the borehole wall, which may be retrieved by a cable linking the gun with the bullet. Typically this tool, consisting of up to 52 shots per gun, is run to acquire samples for geological analysis.

- Sidewall coring: This is an advanced version of the sidewall sampling tool. Instead of firing a bullet into the formation, an assembly is used to drill a sample from the borehole wall, thereby helping to preserve the rock structure for future geological or petrophysical analyses.

- NMR: These tools measure the T_1 and T_2 relaxation times of the formation. Their principles and applicability are described in Chapter 5.

- Vertical seismic profiling (VSP): This tool fires a seismic source at the surface and measures the sound arrivals in the borehole at certain depths using either a hydrophone or anchored three-axis geophone. The data may be used to build a localized high-resolution seismic picture around the borehole. If only the first arrivals are measured, the survey is typically called a well shoot test (WST) or **checkshot** survey. VSPs or WSTs may also be performed in cased hole.

1.2.3 Wireline Cased Hole Logging

When a hole has been cased and a completion string run to produce the well, certain additional types of logging tools may be used for monitoring purposes. These include:

- Thermal decay tool (TDT): This neutron tool works on the same principle as the neutron porosity tool, that is, measuring gamma ray counts when thermal neutrons are captured by the formation. However, instead of measuring the HI, they are specifically designed to measure the neutron capture cross-section, which principally depends on the amount of chlorine present as formation brine. Therefore, if the formation water salinity is accurately known, together with the porosity, S_w may be determined. The tool is particularly useful when run in time-lapse mode to monitor changes in saturation, since many unknowns arising from the borehole and formation properties may be eliminated.
- Gamma ray spectroscopy tool (GST): This tool works on the same principal as the density tool, except that by measuring the contributions arising in various energy windows of the gamma rays arriving at the detectors, the relative proportions of various elements may be determined. In particular, by measuring the relative amounts of carbon and oxygen a (salinity independent), measurement of S_w may be made.
- Production logging: This tool, which operates using a spinner, does not measure any properties of the formation but is capable of determining the flow contributions from various intervals in the formation.
- Cement bond log: This tool is run to evaluate the quality of the cement bond between the casing and the formation. It may also be run in a circumferential mode, where the quality around the borehole is imaged. The quality of the cement bond may affect the quality of other production logging tools, such as TDT or GST.
- Casing collar locator (CCL): This tool is run in order to identify the positions of casing collars and perforated intervals in a well. It produces a trace that gives a "pip" where changes occur in the thickness of the steel.

1.2.4 Pipe-Conveyed Logging

Where the borehole deviation or dogleg severity is such that it is not possible to run tools using conventional wireline techniques, tools are typically run on drillpipe. In essence, this is no different from conventional logging. However, there are a number of important considerations. Because of the need to provide electrical contact with the toolstring, the normal procedure is to run the toolstring in the hole to a certain depth before pumping down a special connector (called a wet-connect) to connect the cable to the tools. Then a side-entry sub (SES) is installed in the drillpipe, which allows the cable to pass from the inside of the pipe to the annulus. The toolstring is then run in farther to the deepest logging point, and logging commences. The reason the SES is not installed when the toolstring is at the surface is partly to save time while running in (and allowing rotation), and also to avoid the wireline extending beyond the last casing shoe in the annulus. If the openhole section is longer than the cased hole section, the logging will need to be performed in more than one stage, with the SES being retrieved and repositioned in the string. Pipe-conveyed logging is expensive in terms of rig time and is typically used nowadays only where it is not possible to acquire the data via LWD.

Most contractors now offer a means to convert an operation to pipe-conveyed logging if a toolstring, run into the hole on conventional wireline, becomes stuck in the hole. This is usually termed "logging while fishing."

1.3 LOGGING CONTRACTS

Typically, an oil company will set up logging contracts with one or more contractors for the provision of logging services. Usually some kind of tendering process is used to ensure competitive bidding among various companies able to provide such services. Elements that exist in common contracts include the following:

- Depth charge: This relates to the deepest depth that a particular tool will be run in the hole.
- Survey charge: This relates to the interval that a particular tool is actually logged in the hole.
- Station charge: For tools such as formation pressure sampling tools and sidewall samples, this is a charge per station measurement. Usually the contract will make certain specifications regarding when such a charge may be dropped (e.g., if no useful data are recovered).
- Tool rental: Usually a daily charge for the tools to be on the rig on standby prior to or during a logging job
- Logging unit rental charge: Usually a monthly charge for the logging unit (winch, tool shed, and computers) while it is on the rig
- Base rental: There may be a monthly charge to have a pool of tools available for a client. For LWD tools, this may supersede the tool rental, depth, and survey charges.
- Engineer charge: Usually a day rate for any engineers, specialists, or assistants present for the logging job
- In-hole charge: Some LWD contracts specify an hourly charge while tools are actually being run in the hole.

- Lost-in-hole charge: For replacement of any tools lost in the hole during operations. Some contractors provide insurance to the oil companies for a fixed sum per job to indemnify them against lost-in-hole charges.
- Cable splice charge: Where tools become stuck in the hole and it is necessary to cut the cable, a charge is usually made for such splicing.
- Processing charges: Where data require postprocessing (e.g., interpretation of image data or waveform sonic), charges are usually applied in a similar way to survey charges.
- Data charges: Provision of additional copies of log prints and/or tapes, or data storage, may incur additional charges.
- Real-time data transmission charges: The oil company will usually be given the option to have data transmitted directly from the wellsite to their office, either as digital data in Log ASCII Standard or binary format or as a print image.

Most contracts offer the oil company a discount on the total monthly charges based on total volume of services called out during a particular month. Some oil companies operate incentive schemes that penalize the contractor financially based on lost time resulting from tool failures. There may also be bonuses based on good safety performance.

When new tools being introduced by the contractor and not covered by the contract are proposed to be run, there will normally be some negotiation on special pricing. The oil company takes into account that if there is a testing element in the new tool being run, there is a benefit accruing to the contractor. Hence, the oil company may argue that the tool should be run free of charge initially. The contractor will usually argue that the oil company is benefiting from the tool's technological advantages over alternative older-generation tools. Often a compromise is reached whereby the tool is run at a preferential pricing scheme (maybe equivalent to the price of the tool being replaced) for the first few runs until its usefulness has been proved. Typically the contractor will request the right to use the data acquired for future promotion of the tool, subject to confidentiality restrictions.

Most oil companies will also specify, either in the contract or in a separate document, how data are to be delivered to their offices and what quality-control procedures should be followed during logging. Items typically specified in such a document will include:

- Pre- and postrun tool calibration procedures
- Sampling increments

- Repeat sections to be performed
- Data items and format to be included in the log header
- Procedures for numbering and splicing different runs in a hole
- Scales to be used in the presentation of logs
- Format and media required for digital data
- Requirements for reporting of time breakdown of logging operation, personnel on site, serial numbers of tools used, inventory of explosives, and radioactive sources
- Specific safety procedures to be followed
- Provision of backup tools
- Fishing equipment to be provided

Generally speaking, the more the oil company specifies its requirements, the better. Having a strict system in place for controlling the logging operation and presentation of results ensures a smooth operation and results in high-quality data that will be consistent with previous runs.

1.4 PREPARING A LOGGING PROGRAM

At the FDP stage in a field development, the outline of the logging strategy should be developed. Based on the type of well being proposed, decisions have to be made about whether to go principally for an LWD type of approach or conventional wireline and about the types of tool to be run.

In general, early in the life of a field, particularly during the exploration phase, data have a high value, since they will be used to quantify the reserves and influence the whole development strategy. Moreover, lack of good-quality data can prove to be extremely expensive, particularly for offshore developments, if facilities are designed that are either too big or too small for the field.

Later in field life, particularly in tail-end production, where much of the log data will not even be used for updating the static model, since it is influenced by depletion effects, the value of data becomes much less. However, even in mature fields, it may be the case that extensions to the main accumulation are still being discovered, and existing assumptions such as those regarding the position of the FWL may need to be locally revised.

The FDP should lay down the broad strategy for data acquisition, which will take into account the relevant uncertainties remaining in the STOIIP and the options for adapting the wider development strategy. It is obviously important that discussion take place between the petrophysicist and the geologist about the need for coring and the various analyses that will be performed once the core is recovered.

For a particular well, the detailed logging requirements will first be specified within the well proposal, which will be agreed upon with other partners and any government supervisory bodies. The proposal will typically not specify the exact models of tool to be run but will cover the general types for the individual hole sections. Typically there will be many items that are conditional on hydrocarbons being encountered, based on shows encountered while drilling. For a well proposal, a typical program might look something like this:

Exploration Well
17¹/₂″ hole section:
GR/resistivity/sonic (GR to surface)
If shows encountered include GR/density/neutron and optional pressure/
  fluid sampling

12¹/₄″ hole section:
LWD GR/resistivity
Wireline GR/resistivity/density/neutron
Optional pressure/fluid sampling if hydrocarbons encountered

8¹/₂″ hole section:
LWD GR/resistivity
GR/resistivity/density/neutron
GR/dipole sonic/formation imager
Pressure/fluid sampling (sampling dependent on oil being encountered)
VSP
Sidewall samples

Development Well
17¹/₂″ hole section:
No logs required

12¹/₄″ hole section:
MWD [measurement while drilling]/GR
GR/resistivity/sonic (GR to surface)
If shows encountered include GR/density/neutron and optional pressure/
  fluid sampling

8$^1\!/_2$" hole section:
GR/resistivity/density/neutron
Dipole sonic/formation imager
Pressure/fluid sampling (sampling dependent on oil being encountered)
VSP
Sidewall samples

Note that it is not usually necessary to specify that the SP will be run, since this service is usually provided free and the log will be included in the first toolstring in the hole by default. Likewise, thermometers are usually run in the toolstring as standard, and the maximum temperature recorded is included in the log header. Prior to the actual logging job in each section, a program is usually sent to the rig with the following more detailed specifications:

- The actual mnemonics of tools to be run (dependent on the contractor)
- Intervals to be logged if different from the total openhole action
- How the tools are to be combined to form the individual toolstrings
- Data transmission/delivery requirements

For the so-called conventional logs (i.e., GR, resistivity, sonic, density, neutron), it is not usually necessary to be very specific, since the company will have already established the tool parameters via generalized guidelines, as discussed in Section 1.3 above. However, the type of resistivity tool to be used will depend on the drilling mud in the hole and the resistivities expected to be encountered. While only induction tools may be run in OBM, the optimum tool to be run in WBM will depend on the ratio of the mud filtrate resistivity (R_{mf}) to the formation, or water, resistivity (R_w). As a rule of thumb, an induction tool is preferred if the ratio of R_{mf} to R_w is more than 2. However, laterolog tools tend to be more accurate in highly resistive formations (resistivity at room temperature >200 ohmm) and are inaccurate below about 1 ohmm. Induction tools, on the other hand, become saturated above 200 ohmm but are more accurate in low-resistive formations. For formation imaging, resistivity tools cannot be used in OBM, although they are definitely preferred in WBM. When OBM is used, it is necessary to use an ultrasonic device.

Usually the stations to be used for pressure/fluid sampling, VSP, and sidewall sampling will be dependent on the analyses made on the first run(s) in the hole. These stations may in some cases be picked on the wellsite by the company's representative but are usually determined in the client's office. Therefore, a second program is usually sent to detail the further logging requirements.

1.5 OPERATIONAL DECISIONS

While the logging program will aim to cover most eventualities during the logging job, often instant decisions have to be made where it is not possible to call everyone into a meeting and get the approval of all parties concerned. Below are some of the things likely to happen and some considerations in decision making.

1.5.1 Tool Failures

The standard procedure if a tool fails is to replace it with a backup and carry on as before. In general, if the program has specified that a tool is to be run over a specified interval (particularly a reservoir interval), then it is best to ensure that good-quality data are acquired, even if it means making additional runs. However, the following situations may arise:

- When logging with LWD and in the hole drilling near the end of a section, it may be far more cost-effective, if the data are not critical, to simply carry on drilling and reacquire the data during a check trip. In some cases, the memory data may still be usable.
- If a wireline tool starts to act erratically, it may not justify rerunning the tool, since the data may not be critical over the particular interval and can sometimes be corrected by postprocessing the raw tool data at the surface.
- If an advanced tool fails, it sometimes happens that no backup tool is available on the rig. The choices then are to either try and repair the tool, arrange a backup tool to be sent from another location, or replace the tool with an earlier version with less capability. Most commonly, the latter option is chosen.
- In the event of failures occurring with potential safety implications (e.g., explosive charges going off accidentally), it is normal for operations to be suspended until a full investigation has been carried out to establish the cause.

1.5.2 Stuck Tools

Fairly regularly during logging operations, tools get stuck in the hole, either temporarily or permanently. There are often indications of bad hole conditions during drilling, and the logging program may be adjusted to take this into consideration. In general it is found that the longer a hole is left open, the greater the likelihood of hole problems occurring. Three sorts of sticking are found to occur:

- differential sticking,
- key seating, and
- holding up.

Differential sticking occurs when either the cable or the toolstring becomes embedded in the borehole wall and gets held in place by the differential pressure between the mud and the formation. In such a situation, it is impossible to move the toolstring either up or down. The usual procedure is to alternately pull and slack off on the tool, pulling up to 90% of the weakpoint of the cable (that point at which the cablehead will shear off the top of the toolstring). Strangely enough, this procedure is often successful, and a tool may become free after 30 minutes or so of cycling.

Key seating occurs when a groove is cut into one side of the borehole, which allows the cable, but not the toolstring, to pass upward. The toolstring is effectively locked in place at a certain depth. Unfortunately this often means that when the weakpoint is broken, the toolstring will drop to the bottom of the hole and may be hard to recover or may be damaged.

Holding up occurs when a constriction, blockage, dogleg, or shelf occurs in the borehole such that the toolstring may not pass a certain depth, although it can be retrieved. The usual practice in such a situation is to pull out of the hole and reconfigure the tool in some way, making the toolstring either shorter or, in some cases, longer in an attempt to work past the holdup depth.

Once it is no longer possible to recover a tool on wireline, there are two options: cutting-and-threading or breaking the weakpoint. In the cut-and-thread technique, the cable is cut at the surface. The drillpipe (with a special fishing head called an overshot) is run into the hole with the cable being laboriously threaded through each stand of pipe. At a certain depth it may be possible to install an SES, which means that when the toolstring is being recovered, log data may be acquired in a similar fashion to pipe-conveyed logging. If the weakpoint has been broken (either accidentally or on purpose), cutting-and-threading is not possible. Then the pipe is run into the hole until the toolstring is tagged, although often the tool will drop to the bottom of the hole before it can be engaged by the overshot.

Most oil companies will specify that they do not wish to break the weakpoint on purpose, even if the cut-and-thread technique is much slower. This is particularly true where nuclear sources have been lost in the hole, and every effort should be made to ensure that they are recovered undamaged at the surface. In the event of nuclear sources being lost irretrievably in the hole, which occasionally happens (particularly with LWD tools), there are usually special procedures to be followed involving notification of government bodies and steps taken to minimize the risk of and monitor any potential nuclear contamination that may occur.

1.6 CORING

1.6.1 Core Acquisition

Particularly during the exploration phase of a field, coring presents an important means to calibrate the petrophysical model and gain additional information about the reservoir not obtainable by logs.

Usually the decision of when and where to core will be made in conjunction with the geologist and operations department, taking into account the costs and data requirements. Generally speaking, it is considered essential to at least attempt to core a part of the main reservoir formation during the exploration and appraisal phases of drilling.

A so-called conventional core will usually consist of multiples of 18 m and be 4 in. in diameter. The outer barrel has a diameter of $6^3/_4$ in. It is acquired while drilling using a metal sleeve into which the core passes during drilling. At the end of coring, the core barrel is retrieved at the surface and the core recovered from the barrel and laid out in 3-ft sections in core boxes for initial assessment on the wellsite and then transportation to the designated core laboratory. Special techniques may sometimes be proposed to improve the quality of the core and to preserve the in-situ fluids. These include:

- Using a large-diameter core (5 in.)
- Using a fiberglass or aluminium inner sleeve, which may be cut into sections at the surface, thereby preserving the core intact within the sleeve
- Sponge coring, whereby a polyurethane material surrounds the core in the inner sleeve, thereby absorbing and retaining any formation fluids
- Resin coring, whereby a special resin is injected onto the surface of the core to seal the fluids inside
- Freezing the core as soon as it reaches the surface in order to preserve the fluids inside

- Cutting plugs from the core at the wellsite, which may be sealed and used to measure the formation fluids
- Using tracers in the mud to attempt to quantify the extent of invasion of drilling fluid

If samples have been obtained and preserved so that it is expected that the in-situ fluids are representative of the formation, the following techniques may be applied:

- Centrifuging of samples to produce formation water, which can be analyzed for chemical composition and electrical properties
- Applying Dean-Stark analysis to determine the relative amounts of water and hydrocarbons, thereby producing a measurement of S_w

1.6.2 Conventional Core Analysis

As soon as possible after drilling, sections of the core (typically 0.5 m every 10 min) are sealed and kept as preserved samples. The remaining whole core is typically cleaned, slabbed, and laid out so that the geologist and petrophysicist can visually inspect the core and examine any sedimentary features. Important information the petrophysicist can learn from such an inspection include:

- The homogeneity of the reservoir and any variations that are likely to be below the resolution of logging tools
- The type of cementation and distribution of porosity and permeability
- The presence of hydrocarbons from smell and appearance under ultraviolet (UV) light. Oil/water contacts (OWCs) can sometimes be established in this way
- The types of minerals present
- Presence of fractures (either cemented, natural, or drilling induced) and their orientation
- Dip features that may influence logging tools' response

After slabbing, the usual procedure is for conventional plugs (typically 0.5 in. diameter) to be cut at regular intervals. The plugs are then cleaned by refluxing with a solvent for 24 hours and dried at a temperature that will remove any water (including clay-bound water). These plugs are then measured for porosity (using a helium porosimeter), horizontal permeability, and grain density. Additional plugs are cut in the axis of the core to determine vertical permeability.

Usually a gamma ray detector or density-type device is run over the whole length of the core in order to provide a reference log that may be correlated to the wireline data. Since the "driller's depths" to which a core is referenced are typically different from "logger's depths," as measured by wireline, it is necessary to make a shift before the core may be compared to logs. The conventional-plug measurements are usually performed at ambient conditions (or sometimes a few hundred psi confining pressure) and therefore need to be corrected to in-situ conditions before they may be compared to the logs. The correction factors to be used are determined through further special core analysis (SCAL).

1.6.3 Special Core Analysis

SCAL measurements are typically performed on a special set of larger-diameter (1.5 in.) plugs cut from the core. These may be cut at a regular sampling increment, or the petrophysicist may specify certain depths based on the results of the conventional analyses. The most important criterion is obviously to obtain a broad spectrum of properties that fully encompass the range of properties seen in the reservoir.

In order to ensure that the SCAL plugs are homogeneous, it is normal procedure to subject the plugs to a CAT (computed axial tomography) scan prior to using them for future measurements. It is hard to say how many SCAL plugs are required for a typical program, since this depends on the reservoir type, thickness, and homogeneity. In general a SCAL program may use between about 5 and 50 plugs.

While many measurements are possible on core plugs, I will concentrate on the ones that are of direct relevance to the petrophysical model. These are:

- **Porosity and permeability at overburden conditions.** Here it is important to state the pressures at which the measurements should be performed. In Chapter 7 the equations are given for calculating the equivalent isostatic stress at which the measurements should be performed to be equivalent to in-situ conditions. Typically measurements are made at five pressures that will encompass the likely range of pressures to be encountered during depletion of the reservoir.
- **Cementation exponent (*m*).** In this measurement, the resistivity of the plugs is measured when they are 100% saturated with brine represen-

tative of the formation salinity. This measurement is usually performed at ambient conditions but may also be performed at in-situ pressure.

- **Saturation exponent (*n*).** In this measurement, the resistivity of the plugs is measured as a function of water saturation, with the resistive fluid being either air or kerosene. This measurement is usually performed at ambient conditions.
- **Capillary pressure (P_c).** The saturation of a nonwetting fluid (either air, mercury, or kerosene) is measured as a function of P_c applied. In a drainage cycle, 100% brine is gradually replaced by the nonwetting fluid. For an imbibition cycle (following a drainage cycle), brine is reintroduced to replace the nonwetting phase.

Different techniques are available to make these measurements. In the traditional approach, *m*, *n*, and P_c would be measured using the porous plate method, with air as the nonwetting phase. Since the measurement is limited to 100 psi, additional P_c measurements would be performed using mercury injection up to 60,000 psi, thereby also determining the pore-size distribution.

Many oil companies no longer favor these measurement techniques for the following reasons:

1. Measurements using mercury involve destruction of the plugs and present a potential environmental/health hazard.
2. P_c measurements involving air/mercury are not representative of true reservoir conditions and may give misleading results.
3. Porous plate measurements are slow and involve the repetitive handling of the samples to measure the saturations using a balance. If grain loss occurs, then the results are inaccurate and the electrical measurements tend to be operator dependent.

Preferred techniques for undertaking these measurements are as follows:

- Measurement of *m* and *n* should be performed using a continuous injection apparatus. While not steady state, this technique has been shown to give reliable results. In the procedure, the sample is mounted vertically, flushed with brine, then kerosene-injected at a continuous rate while the resistivity and saturation are continually monitored.
- P_c should be measured using a centrifuge capable of up to 200 psi pressure. The sample is flushed with brine, then the amount of fluid expelled

at different rotational speeds (equivalent to different pressures) is measured. This technique also has the advantage that the sample if not handled during the experiment.

1.6.4 Limitations of Core Measurements

There is a tendency among petrophysicists to treat measurements made on cores as "gospel" and not to question the reservoir parameters so derived in their petrophysical model. The following may give reasons why the core data are not always correct:

- A core is a section of rock cut usually over only a subset of the reservoir in a particular part of a field. There is no *a priori* reason why it should be representative of the reservoir as a whole. In particular, a core cut in the water leg, where diagenetic processes may be occurring, is not necessarily representative of the oil or gas legs in a reservoir.
- The coring and recovery process subjects the rock to stress and temperature changes that may profoundly affect the rock structure.
- The plugging, cleaning, and drying process may completely change the wettability of the plugs, making them unrepresentative of downhole conditions.
- Resistivity measurements performed on plugs at ambient temperature, using air as the nonwetting fluid, may be wholly unrepresentative of reservoir conditions. Apart from the fact that the brine has a totally different resistivity at ambient temperatures, there may be other factors affecting how easily the nonwetting phase may mingle with the wetting phase. In fact, where experiments have been performed to measure m and n under truly in-situ conditions, it was found that the values differed completely from those measured under ambient conditions.
- When measurements are made on a selection of, say, 10 SCAL plugs, it will typically be found that the m, n, and P_c behavior of all 10 will be completely different. These are usually then averaged to obtain a representative behavior for the reservoir. However, because of the variability, if a new set of 10 plugs is averaged, the result will be completely different. This calls into question the validity of any average drawn from 10 plugs that are taken to represent thousands of acre-feet of reservoir.

Overall, it is my conclusion that it is better to use core-derived values than nothing at all, and a lot of valuable information about the reservoir can be gained from core inspection. However, no core-derived average should be treated as being completely reliable, and there will be many cases in which it has to be disregarded in favor of a commonsense approach to all the other sources of information.

1.7 WELLSITE MUD LOGGING

During the drilling of a well there will typically be a mud-logging unit on the rig. This unit has two main responsibilities:

1. To monitor the drilling of the parameters and gas/liquids/solids returns from the well to assist the drilling department in the safety and optimization of the drilling process
2. To provide information to the petroleum engineering department that can be used for evaluation purposes

Typically the mud-logging unit will produce a daily "mud log," which is transmitted to the oil company office on a daily basis. Items that will be included are:

- Gas readings as measured by a gas detector/chromatograph
- A check for absence of poisonous gases (H_2S, SO_2)
- A report of cuttings received over the shale shakers, with full lithological descriptions and relative percentages
- ROP
- Hydrocarbon indications in samples

The mud log may be of great use to the petrophysicist and geologist in operational decision making and evaluation. Areas in which the mud log may be particularly important include:

- Identification of the lithology and formation type being drilled
- Identification of porous/permeable zones
- Picking of coring, casing, or final drilling depths
- Confirmation of hydrocarbons being encountered and whether they are oil or gas

1.7.1 Cuttings Descriptions

The mud-logging unit will generally take a sample of the cuttings received over the shale shakers at regular time intervals, calculated to correspond to regular changes in formation depth (e.g., every 5 m). Some of these samples are placed into sealed polythene bags as "wet samples" and retained. Other samples are washed, dried, and retained as "dry samples." Washed samples are examined under a microscope in the mud-logging unit and a description made that may be communicated to the office.

In order for the information received from the rig to be useful, it is essential that rigid standards for reporting are followed that are agreed upon between the rig and the office. Standards will typically vary among companies. Items that should be included are:

- Grain properties
  - Texture (muddy/composite)
  - Type (pelletoid/micropelletoid)
  - Color
  - Roundness, or sphericity
  - Sorting
  - Hardness
  - Size
  - Additional trace minerals (e.g., pyrite, calcite, dolomite, siderite)
  - Carbonate particle types
  - Skeletal particles (fossils, foraminifera)
  - Nonskeletal particles (lithoclasts, aggregates, rounded particles)
  - Coated particles
- Porosity and permeability
  - Porosity type (intergranular, fracture, vuggy)
  - Permeability (qualitative as tight, slightly permeable, highly permeable)
- Hydrocarbon detection

Hydrocarbons may be detected with one of the following methods:

Natural fluorescence

Examining the cuttings under UV light may indicate the presence of oil, since oil will fluoresce. However, fluorescence will not in itself prove the presence of movable oil, due to other sources of fluorescence that may be present, such as fluorescent minerals; OBM or lubricants used; other sources of carbon, such as dead oil or bitumen; and Gilsonite cement.

The correct procedure is for a portion of the lightly washed and undried cuttings to be placed on a dish and observed under UV light (other light sources having been removed). Those parts of the sample exhibiting fluorescence are picked out and placed in a porcelain test plate hole to be examined for cut fluorescence.

Solvent cut

To measure the solvent cut, about 3 cm of dried and crushed sample is placed in a test tube and solvent is added to about 1 cm above the sample. The test tube is shaken for a few minutes, then left to stand. The solvent cut is the change of coloration of the solvent. Solvents that are commonly used are chlorothene, ether, and chloroform. Precautions are required in handling these solvents, since they are toxic and flammable. Heavy oils generally give a stronger cut than lighter ones. Asphalts will therefore give a stronger cut than paraffins. Condensate gives rise to only a very light cut. In addition to the cut, a residual oil ring may be observed around the test tube after the solvent has evaporated.

In solvent cut fluorescence, the cut fluorescence is measured by taking the test tube used for the solvent cut and placing it under UV light together with a sample of the pure solvent (to check for possible contamination) and observing whether any fluorescence is present.

Acetone test

The acetone test involves placing a sample of washed, dried, and crushed cuttings in a test tube with acetone. After shaking, the acetone is filtered into another test tube and an equal amount of water added. Since acetone is dissolvable in water but hydrocarbons are not, the liquid becomes milky in color. This test is particularly useful where light oil or condensate is present and there is no other source of carbon in the samples.

Visible staining

Particularly if the permeability and/or viscosity is poor, oil may remain in cuttings and be visible under the microscope in the form of a stain on the surface of the cutting.

Odor

The characteristic smell of oil may sometimes be discerned during the cleaning and drying process.

Gas detection analysis

Gas detectors work by passing air drawn from where the mud reaches the surface (the bell nipple) over a hot detector filament. This combusts the gas, raising the temperature and lowering the resistance of the filament. At high voltages all the combustible gases burn, whereas at lower temperatures only the lighter components burn. By recording the change in resistance at different voltages, the relative proportions of the various components may be estimated.

A gas chromatograph may also be used to further differentiate the various hydrocarbon components. Particularly for the detection of poisonous gases, such as H_2S, Drager tubes may be used on the rig floor.

1.8 TESTING/PRODUCTION ISSUES

At the end of the logging, a decision will have to be made as to whether casing should be run or not. If the well results are not as expected, there may be an immediate decision required to either sidetrack or abandon the well. Therefore, a quick but accurate interpretation of the data, not always made using a computer, is of primary importance.

If the decision is made to test or complete the well, the petrophysicist will also be required to pick the perforation intervals. A few points to bear in mind here are that when picking the intervals from a log it is important to specify the exact log being used as a depth reference. Since depths on field prints are sometimes adjusted to tie in with previous runs when the final prints are made, confusion can occur. The safest thing is to include a photocopy of the reference log, with the intervals to be perforated marked on the log, along with any program passed to the rig.

The correct procedure for ensuring that the well is perforated "on depth" with wireline operations is as follows:

1. Initially it is necessary to tie the depths of casing collars, as measured in the well using a CCL, with the reference openhole GR log. This is done by making a run in the hole with a GR/CCL tool and comparing the depth with the openhole reference log on which the perforated interval has been marked.
2. The print of the GR is then adjusted so that the depths of the casing collars are on depth with the openhole GR.

3. When the perforating guns are run, they will be combined with only a CCL log. On first running in the hole, the CCL on the perforating guns will be off depth with the corrected GR/CCL.

4. Because of irregularities in the length of joints of casing, the CCL acquired with the guns may be overlain on the GR/CCL and a unique fit made. This enables the logger's depths of the perforating guns to be adjusted so that the CCL is on depth with the GR/CCL (which is itself on depth with the reference openhole log). Obviously if all the joints of casing were the same length, it would be possible to find a fit when the toolstring was off depth by the length of one casing joint. This problem may be avoided by running a radioactive pip tag as part of the completion, which enables the CCL on the perforating gun to be tied with certainty to the GR/CCL.

5. Once the gun has been fired, there may be indications on the surface, such as changes in cable tension. After a few minutes, there may also be indications of an increase in tubing head pressure.

The most appropriate guns, charges, phasing, and well conditions (fluid type and drawdown) all need to be considered. Usually the contractor is able to offer advice on this and should be involved in any meetings at which the perforation procedures are to be determined. Wherever possible it is best that wells be perforated "under drawdown." This means that the pressure in the wellbore is lower than the formation pressure, which ensures that the well is able to flow as soon as perforation has occurred and avoids the risk of either completion fluid or debris blocking the perforations.

I have seen many cases in which the petrophysicist has picked many short intervals to be perforated, separated by only a foot or so. Since the accuracy of depth correlation is never perfect, it is sometimes advisable to shoot a continuous section, which includes parts that are not of the reservoir. There has always been caution about perforating shales, in case they result in "fines" being produced. I can only say that I have never heard of this occurring in practice. In general I favor perforating as much of the potentially producible interval as possible (a safe distance from any water-bearing sands). If you look at the economics of a well, an additional 10 bbl/day over the life of a well will result in a far greater economic benefit than the additional cost of perforating an extra 10 m. In some cases, picking too short an interval may result in the well never even managing to flow to surface when otherwise it could be an economic producer. It may frequently occur, particularly in depleted reservoirs, that the well doesn't produce as it was supposed to, as per calculations. Either it fails to produce at all, it dies quickly, or water or gas breakthrough occurs. Very often the petrophysicist is called to provide an explanation for these phenomena. When a well fails to flow or dies very quickly, the first thing to look at is the perforation operation. Check the following:

- Is the petrophysical interpretation completely reliable? What was the k_h as derived from the logs?
- Could formation damage have occurred during the cementation and completion operation?
- Is there proof that the guns detonated, and were they on depth?
- What was the over/underbalance at the time of perforation? What drawdown is currently being applied?
- Are there any other mechanical factors (sliding side doors, safety valves) that could prevent flow?

When water breakthrough occurs sooner than expected, the following should be considered:

- How close are the perforations to any water leg as logged in the well? Could water coning be occurring?
- What is the quality of the cement bond, and could there be flow behind casing?
- Where does the dynamic model of the reservoir predict the encroaching water front to be?
- Are neighboring wells already starting to water out, and are TDT/GST data available for any?
- Could the water be entering from elsewhere in the wellbore (other producing zones, leaks in the tubing, etc.)?
- What does the relative permeability data from the core indicate the relative permeability to oil and water to be?
- Is the well in the transition zone, and how sure are you that any S_w calculated on the logs is capillary or clay bound?

When gas is produced unexpectedly, the following should be considered:

- How close are the perforations to any gas leg as logged in the well? Could gas cusping be occurring?
- What is the quality of the cement bond, and could there be flow behind the casing?

- Where does the dynamic model of the reservoir predict the GOC to be?
- Is the bottomhole pressure below the bubble point? What choke size is being used?
- Are neighboring wells already starting to produce gas?
- Could the gas be entering from elsewhere in the wellbore (other producing zones, leaks in the tubing, etc.)?
- What does the relative permeability data from the core indicate the relative permeability to oil and gas to be?

Once the above questions have been answered for the various scenarios, the petrophysicist will be in a good position to contribute to discussions regarding remedial actions to be taken. Such actions may include any of the following:

- Further data gathering through the use of production logs
- Reperforation
- Acidization
- Acid or hydraulic fracturing of the reservoir
- Sealing off of certain zones either chemically or mechanically
- Modification of the offtake strategy
- Recompletion
- Sidetracking of the well
- Implementation of artificial lift techniques (e.g., gas lift or ESP [electric submersible pump])
- Implementation of a water or gas injection program

CHAPTER 2

QUICKLOOK LOG INTERPRETATION

Once the section TD (total depth) of the hole has been reached, the petrophysicist will be expected to make an interpretation of the openhole logs that have been acquired. Before starting the log interpretation, the petrophysicist should have:

1. All the relevant daily drilling reports, including the latest deviation data from the well, last casing depth, and mud data
2. All the latest mud-log information, including cuttings description, shows, gas reading, and ROP (rate of penetration)
3. Logs and interpretations on hand from nearby wells and regional wells penetrating the same formations, in particular where regional or field-wide values of m, n, R_w, rho_g and fluid contacts are available
4. A copy of the contractor's chart book

2.1 BASIC QUALITY CONTROL

Once the log arrives, the petrophysicist needs to ensure the quality of the log data and should perform the following regimen:

1. Check that the logger's TD and last casing shoe depths roughly match those from the last daily drilling report.
2. Check that the derrick floor elevation and ground level (or seabed) positions are correct.
3. Check that the log curves are on depth with each other. The tension curve can be used to identify possible zones where the toolstring has become temporarily stuck, which will put the curves off depth and result in "flatlining."

4. Check that the caliper is reading correctly inside the casing (find out the casing ID) and that it is reading the borehole size in nonpermeable zones that are not washed out.
5. Check the density borehole correction curve. It should not generally exceed 0.02 g/cc, except in clearly washed out sections (>18 in.), for which the density curve is likely to be unusable.
6. Inspect the resistivity curves. If oil-based mud (OBM) is being used, the shallow curves will usually read higher than the deep curves (except in highly gas or oil saturated zones). Likewise, with water-based mud (WBM) the shallow curves will read less than the deep curves, providing $R_{mf} < R_w$, or in hydrocarbon-bearing zones. In theory, the curves should overlie each other in nonpermeable zones such as shales. However, in practice this is often not the case, due to either anisotropy or shoulder-bed effects.
7. Check the sonic log by observing the transit time in the casing, which should read 47 μs/ft.
8. Look out for any cycling-type behavior on any of the curves, such as a wave pattern. This may be due to corkscrewing while drilling, causing an irregular borehole shape. However, it is necessary to eliminate any possible tool malfunction.
9. Check that the presentation scales on the log print are consistent with other wells or generally accepted industry norms. These are generally:
   - GR: 0–50 API
   - Caliper: 8–18″
   - Resistivity: 0.2–2000 ohmm on log scale
   - Density: 1.95–2.95 g/cc (solid line)
   - Neutron: −0.15 ± 0.45 (porosity fraction) (dashed line)
   - Sonic: 140–40 μs/ft

2.2 IDENTIFYING THE RESERVOIR

For the next section of this chapter, it will be assumed that one is dealing with clastic reservoirs. Carbonates and complex lithologies will be discussed later in the book. The most reliable indicator of reservoir rock will be from the behavior of the density/neutron logs, with the density moving to the left (lower density) and touching or crossing the neutron curve. In clastic reservoirs in nearly all cases this will correspond to a fall in the gamma ray (GR) log. In a few reservoirs, the GR is not a reliable indicator of sand, due to the presence in sands of radioactive minerals. Shales can be clearly identified as zones where the density lies to the right of the neutron, typically by 6 or more neutron porosity units.

The greater the crossover between the density and neutron logs, the better the quality of the reservoir. However, gas zones will exhibit a greater crossover for a given porosity than oil or water zones. Because both the neutron and density logs are statistical measurements (i.e., they rely on random arrivals of gamma rays in detectors), they will "wiggle" even in completely homogeneous formations. Therefore, it is dangerous to make a hard rule that the density curve must cross the neutron curve for the formation to be designated as net sand. For most reservoirs, the following approach is safer (see Figure 2.2.1):

1. Determine an average GR reading in clean sands (GR_{sa}) and a value for shales (GR_{sh}). For GR_{sh}, do not take the highest reading observed, but rather the mode of the values observed.

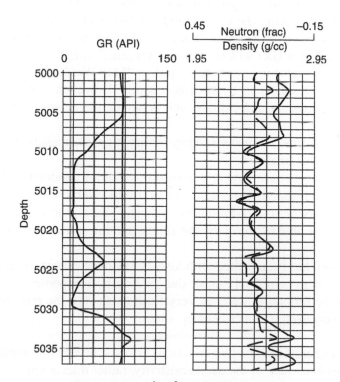

Figure 2.2.1 Identifying Net Reservoir

2. Define the shale volume, V_{sh}, as $(GR - GR_{sa})/(GR_{sh} - GR_{sa})$. By comparing V_{sh} with the density/neutron response, determine a value of V_{sh} to use as a cutoff. Typically 50% is used.

If the GR is not usable as a sand indicator, then for now just treat the entire gross as being net sand and apply a porosity cutoff at a later stage (see next section).

2.3 IDENTIFYING THE FLUID TYPE AND CONTACTS

Because the porosity calculation will depend on the formation fluid type, it is good at this stage to at least have a working assumption regarding the fluids. If regional information is available regarding the positions of any gas/oil contact (GOC) or oil/water contact (OWC), then convert these subsea depths into measured depths in the current well and mark them on the logs. If the formation pressures have already been measured (this is usually never the case), then any information on possible free water levels (FWLs) or GOCs can also be marked on the log.

Start by comparing the density and deepest reading resistivity log for any evidence of hydrocarbons. In the classic response, the resistivity and density (and also GR) will be seen to "tramline" (i.e., follow each other to the left or right) in water sands and to "Mae West" (i.e., be a mirror image of each other) in hydrocarbon sands. However, some hydrocarbon/water zones will not exhibit such behavior, the reasons being:

- When the formation-water salinity is very high, the resistivity may also drop in clean sands.
- In shaly sand zones having a high proportion of conductive dispersed shales, the resistivity may also fail to rise in reservoir zones.
- If the sands are thinly laminated between shales, the deep resistivity may not be able to "resolve" the sands, and the resistivity may remain low.
- If the well has been drilled with very heavy overbalance, invasion may be such as to completely mask the hydrocarbon response.
- When the formation water is very fresh (high R_w), the resistivity may Mae West even in water-bearing zones.

When either of the first two situations arises, it is very important to look at the absolute value of the deep resistivity, rather than at only the behavior compared with the density. As long as a known water sand has been penetrated in the well (or a neighboring well), one should already have a good idea of what the resistivity ought to be for a water-bearing sand. If the resistivity is higher than this value, whatever the shape of the curve, then hydrocarbons should be suspected.

Obviously any mud-log data (gas shows, fluorescence) should be examined in the event that it is not clear whether or not the formation is hydrocarbon bearing. However, the mud log can certainly not be relied on to always pick up hydrocarbons, particularly where the sands are thin and the overbalance is high. Moreover, some minor gas peaks may be observed even in sands that are water bearing (Figure 2.2.2).

As stated earlier, gas zones will exhibit a greater density/neutron crossover than oil zones. In a very clean porous sand, any GOC can be identified on the log relatively easily. However, in general, GOCs will be identified correctly in only about 50% of cases. Secondary gas caps appearing in depleted reservoirs will usually never be picked up in this way. Formation-pressure plots represent a much more reliable way to

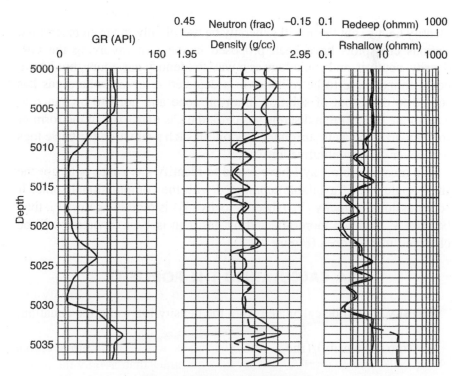

Figure 2.2.2 Identifying Net Pay

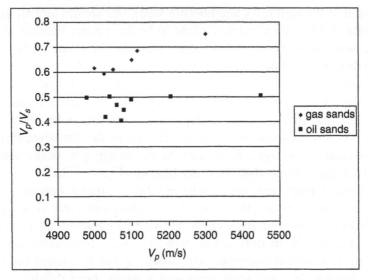

Figure 2.2.3 Identifying Gas Sands from V_p/V_s vs. V_p plot

identify GOCs, but these will generally be useful only in virgin reservoirs. Various crossplots have been proposed in the past involving the GR, density, neutron, and sonic logs as a way to identify gas zones, but I have never found these to be reliable. In a depleted reservoir where gas has started to come out of solution in an oil zone and not had a chance to equilibrate (i.e., form a discrete gas cap), the gas may exist in the form of football-sized pockets surrounded by oil. In such a situation the basic logs will never give a definitive answer.

The most reliable way I have found to identify gas zones is to use the shear sonic log (if available) combined with the compressional sonic. If compressional velocity (V_p)/sheer velocity (V_s) is plotted against V_p, then because V_p is much more affected by gas than is V_s, a deviation will be observed in gas zones (Figure 2.2.3).

2.4 CALCULATING THE POROSITY

Porosity should be calculated from the density log using the equation:

$$\phi = (\text{rho}_m - \text{density})/(\text{rho}_m - \text{rho}_f)$$

where rho_m = matrix density (in g/cc) and rho_f = fluid density (in g/cc).

Table 2.4.1
Selection of fluid density for porosity calculated from density tool

Formation Fluid	OBM	WBM	
		Heavy Mud System	Light Mud System
Gas, clear gas effect on logs	0.4	0.6	0.5
Gas, no clear gas effect on logs	0.55	0.7	0.6
Light oil	0.6	0.8	0.7
Heavy oil	0.7	0.9	0.8
Low-salinity water	0.85	1.05	1.0
High-salinity water	0.9	1.1	1.05

The density tool actually works by injecting gamma rays into the formation that are then scattered by electrons in the formation, a process known as Compton scattering. These gamma rays are then detected by two detectors. Since the tool actually measures electron density, there is a slight miscalibration due to the variation in electron density between different minerals. The correction is typically small (typically 1% or less), so is no major cause for concern. Assuming that the density porosities will at some stage be calibrated against core data, this correction can be ignored, at least for quicklook purposes.

For sandstones, rho_m typically lies between 2.65 and 2.67 g/cc. Where regional core data are available, the value can be taken from the average measured on conventional core plugs. Fluid density, rho_f, depends on the mud type, formation fluid properties, and extent of invasion seen by the density log. Table 2.4.1 gives some typical values that may be used.

As to the appropriateness of the values being used, the following tests may be applied:

- Where regional information is available, the average zonal porosities may be compared with offset wells.
- In most cases, there should be no jump in porosity observed across a contact. An exception may occur across an OWC where diagenetic effects are known to be occurring.
- In no cases in sandstones would one expect porosities to exceed about 36%.

Note that the porosity calculated from the density log is a total porosity value; that is, water bound to clays or held in clay porosity is included.

This has the advantage, therefore, of being directly comparable to porosities measured on core plugs, since these have had all clay-bound and free water removed.

Having calculated the porosity, it is important to check for any zones where washouts have resulted in erroneously high density values and thus unrealistically high porosities. In some cases it is sufficient to just apply a cutoff to the data whereby porosities above a certain value are capped at a value. This recognizes the fact that zones often wash out because they are soft and have a high porosity. However, in some cases it is necessary to manually edit the density log using one's best estimate of what the density should be. Note that in water-bearing sections a good estimate of porosity, ϕ, may be made using true resistivity (R_t) and Archie's equation, which is:

$$R_t = R_w * \phi^{-m} * S_w^{-n}$$

or

$$S_w = [(R_t/R_w) * \phi^m]^{(-1/n)}.$$

where R_w = formation water resistivity (measured in ohmm)
$\quad\quad m$ = the cementation, or porosity, exponent
$\quad\quad S_w$ = water saturation
$\quad\quad n$ = saturation exponent.

Alternatively, sometimes a correlation can be made between the GR and density in non–washed-out zones and applied.

I generally favor always working in a total porosity system. The term *effective porosity* is also used, although often different people take it to mean different things. Probably the best definition is that it is the total porosity minus the clay-bound water and water held as porosity within the clays. It may therefore be defined as:

$$\phi_{eff} = \phi_{total} * (1 - C * V_{sh})$$

where C is a factor that will depend on the shale porosity and CEC (cation exchange capacity). It may be determined from calculating the total porosity in pure shales $(V_{sh} = 1)$ and setting ϕ_{eff} to zero. However, I have doubts about the correctness of assuming that properties of the shales in non-reservoir zones can be applied to dispersed shales within sands in the reservoir. In general I do not recommend calculating ϕ_{eff} at all as part of any quicklook evaluation. At this point I would like to make it clear that I never favor making use of the neutron/density crossplot log for calculating porosity in sandstones. My reasons for this are as follows:

1. Both the neutron and density logs are statistical devices and vary randomly within certain limits determined by the logging speed, detector physics, source strength, and borehole effects. The error introduced when two such random devices are compared is much higher than when one such device is used on its own.
2. The neutron is severely influenced by the amount of chlorine atoms in the formation, occurring either in the formation water or in the clay minerals. This means that the neutron porosity is only very loosely related to the true porosity (as observed when it is compared with the density log in sand/shale sequences!).
3. The neutron is also affected in an unpredictable way by gas (unlike the density, for which a correction can be made using the appropriate rho_f).
4. I have never had much faith in the overlays presented on standard neutron/density crossplots by the contractors. In practice, when real data are plotted, the overlays typically predict all kinds of minerals, from dolomite to limestone, to be present when in fact one is dealing with a clay/quartz combination.

When I do a quicklook I use the neutron log for only two things: (1) qualitative identification (using the density) of shale/sand zones and (2) identification of gas zones. I also do not favor the use of the sonic log for porosity determination under any circumstances. In my view you are better off just making an informed guess at the formation porosity based on the general log response and regional information, rather than relying on any quantitative calculation based on the compressional sonic.

2.5 CALCULATING HYDROCARBON SATURATION

In most quicklook evaluations of clastic reservoirs, it is sufficient to use Archie's equation (see above) to calculate saturations, using the deepest reading resistivity tool directly as R_t. In the absence of any regional core values, I would recommend using $m = n = 2$. Note that I have not chosen to include the so-called Humble constant (a), since this can just as easily be incorporated within R_w.

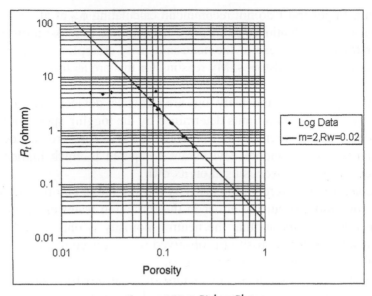

Figure 2.5.1 Pickett Plot

If *m* and *n* are predefined, then clearly the key parameter that must be determined is R_w. By far the preferred method of determining the best R_w to use in a particular well evaluation is a Pickett plot over a known water-bearing section of the formation (Figure 2.5.1). By plotting Log(R_t) vs. log(ϕ), *m* may be determined from the gradient of the line drawn through the points, and R_w may be read from the intercept of the line with the R_t axis.

Note that if *m* is fixed, the line can be moved only up and down. At this point, if the slope of the data is clearly at odds with the assumed *m* value, I would recommend changing *m*, provided that it still lies within a reasonable range (1.5–2.5).

Some information regarding R_w may also be available from regional data and produced water samples in neighboring wells. Note that this will usually be in the form of a salinity expressed as NaCl concentration in ppm or mg/l. This has to be converted to an R_w value using the contractor's chart book and knowledge of the formation temperature.

Where no clear water legs have been logged in the well, there is no alternative but to use regional data, although the Pickett plot may yield a different value than that expected from regional information. Reasons for this may be one of the following:

- The porosities calculated in the well are incorrect.
- The zone may not in fact be 100% water bearing as assumed.
- The value of m needs to be adjusted.
- The regional value is not applicable in this well.

Reasons why the regional value may not apply are:

- The salinity may be different in this well.
- The chart books assume that the conductivity of the brine is caused only by the presence of NaCl. If other chlorides are present (e.g., MgCl), the R_w calculated from the chart book will be wrong.
- The samples from which the salinity has been measured in other wells may be contaminated or affected by salt dropout when the samples were recovered at surface.
- If shale effects are predominant, the conductivity arising from clay-bound water may have a different salinity from that produced in a well. Typically, clay-bound water will be fresher than free water.
- The water zone may have originally been oil bearing but became flushed by injection water of a different salinity (this is common off-shore, where seawater is often used for injection).

In theory, the spontaneous potential (SP) curve may be used in some instances to derive a measurement of R_w, although I have never had much success with it. The procedure for doing this is as follows:

1. On the SP log, draw a shale baseline, which is a line defining an average of the SP readings in 100% shale zones.
2. Determine the maximum SP deflection (in mV) from the baseline to the reading observed in thick, porous parts of the reservoir.
3. Using the appropriate chart as supplied by the contractor, convert the maximum SP deflection to a static spontaneous potential (SSP) value. This corrects for invasion, borehole, and bed effects.
4. From the appropriate chart, determine kinetic energy (E_k), mc (the mudcake contribution).
5. Calculate E_k, shale using E_k, shale $= \Delta P(\text{bar})/6.9$, where ΔP is the pressure difference between the mud pressure and formation pressure.
6. Calculate the Eckert number (Ec) of the bottomhole temperature (BHT) in °C:

$$Ec(BHT) = SSP + E_k, mc - E_{ksh}.$$

7. Correct Ec(BHT) to standard temperature using:

$$Ec(25°C) = Ec(BHT) * 298/(273 + BHT).$$

8. Determine the mud filtrate salinity from R_{mf} and temperature.
9. Determine Q_{vshale} using the appropriate chart as supplied by the contractor. Or else use $Q_{vshale} = 4$ mmol/cc.
10. With the appropriate chart, use Ec(25°C), mud salinity, and Q_{vshale} to determine the formation-water salinity.
11. Convert the formation-water salinity to R_w using the BHT.

At the end of the day it is essential that the model used calculates 100% water in known virgin water-bearing reservoirs. If this is not the case, you may be certain that the S_w calculated in the reservoir will be incorrect.

2.6 PRESENTING THE RESULTS

Having calculated the ϕ and S_w curves, it is usually required to provide averages over various formation zones. This should be done as follows. First of all determine over which depths the results should be broken up. Apart from the formation boundaries as agreed upon with the geologist, further subdivision should be made for any possible changes in fluid type or zones where the data are of particularly poor quality, or at any points where there is marked change in log character. A table such as Table 2.6.1 should be produced.

Note that the average porosity is given by:

$$(\phi)\text{average} = \Sigma \phi_i /h, \text{ where } h \text{ is the net thickness.} \qquad (2.6.1)$$

The average value of S_w is given by

$$(S_w)\text{average} = \Sigma \phi_i * S_{wi} /\Sigma \phi_i. \qquad (2.6.2)$$

Where a permeability transform is available, the average permeability over each major sand body should also be presented.

Usually the net may be defined on the basis of a V_{sh} cutoff. However, where this has not been possible, a porosity cutoff should be used. Generally, the cutoff point should be set at a value equivalent to a per-

Table 2.6.1
Reporting the results of an evaluation

Zone	Top (m)	Base (m)	Gross (m)	Net (m)	Average Porosity	S_w
Zone 1, gas						
Zone 1, oil						
Zone 1, possible oil						
Zone 1, water						
Zone 2						
Zone 3						
Total gas zones						
Total oil zones						

meability of 1 millidarcy (md) for oil zones and 0.1 md for gas zones. In general I don't favor the idea of cutoffs, because all too often they result in potential reserves being excluded from the calculation of STOIIP (stock tank oil initially in place) or GIIP (gas initially in place). However, since Archie's equation will often yield nonzero hydrocarbon saturations in 100% nonreservoir shales, it is usually necessary to apply some kind of cutoff to the data.

I particularly object to the practice of applying a further S_w cutoff and deriving a "pay" footage for a zone. Such a number has no place whatsoever in any kind of STOIIP or GIIP calculation. In theory, a pay footage might be used to assist in decision making regarding which zones to perforate. However, in practice this is performed more effectively by laying out a 1:200 print of the evaluated logs and deciding on that basis which zones are worth perforating. For presentation purposes it is useful to generate a 1:500 version of the evaluated log, with as much data included as possible. Although different companies use different conventions, it is common to use green for gas, yellow for unidentified hydrocarbon, red for oil, and blue for water zones.

I would recommend generating a curve called *SHPOR*, derived from $(1 - S_w)*$Por, and include it in the porosity track, shading from 0 to the curve using the appropriate fluid color. This curve is useful because the area colored is representative of the total volume of the fluid. Hence a thin zone having a high porosity is given more prominence than a thicker zone that might have a much lower porosity.

Exercise 2.1 Quicklook Exercise

Using the log data presented in Appendix 1 (test1 well), do the following:

1. Pick GR_{sa} and GR_{sh} from inspection of the logs.
2. Calculate V_{sh}.
3. Pick the likely position of the OWC.
4. Assuming appropriate fluid densities for the oil and water legs (well was drilled with fresh WBM) and a grain density of 2.66 g/cc, calculate the porosity.
5. Set the porosity to zero wherever $V_{sh} > 0.5$.
6. Make a Pickett plot over the water-bearing interval.
7. Assuming that $m = n = 2$, choose an appropriate R_w.
8. Calculate S_w using Archie's equation.
9. Check the assumed position of the OWC. If it needs to be moved, recalculate the porosity (and S_w) accordingly.
10. Calculate *SHPOR*. Display *SHPOR* in the same depth track as the porosity.
11. Divide the formation into appropriate intervals and calculate sums and averages.
12. Suggest points for the formation pressure tool.

2.7 PRESSURE/SAMPLING

In most cases there will be a requirement to run the pressure/sampling tool to acquire pretests and possibly downhole samples. While these data are also used by the reservoir engineer and production technologist, they can be extremely valuable to the petrophysicist in determining the fluids present in the formation.

Pretests can provide the following information:

- The depths of any FWLs or GOC in the well
- The in-situ fluid densities of the gas, oil, and water legs
- The absolute value of the aquifer pressure and formation pressure
- A qualitative indication of mobility and permeability
- The bottomhole pressure and temperature in the wellbore

Additionally, acquiring downhole samples can provide the following information:

- Pressure/volume/temperature (PVT) properties of the oil and gas in the reservoir
- Formation-water salinity
- Additional mobility/permeability information

In the conventional mode of operation, a probe is mechanically forced into the borehole wall and chambers opened in the tool into which the formation flows. Pretest chambers are small chambers of a few cubic centimeters that can be reemptied before the next pretest station. For downhole sampling, larger chambers are used, typically $2^3/_4$ or 6 gallons. Since the first fluid entering the tool is typically contaminated by mud filtrate, normal practice is to make a segregated sample; that is, fill one chamber, seal it, and then fill a second chamber (hopefully uncontaminated). Once the chambers are retrieved at surface, they may be either drained on the wellsite or kept sealed for transferring to a PVT laboratory.

Optional extra modes in which the tools can typically be used include the following:

1. As an arrangement of packers in order to isolate a few meters of the borehole wall, thereby providing a greater flow area
2. As a pumpout sub while sampling in order to vent the produced fluids into the wellbore until it is hoped that the flow is uncontaminated by mud filtrate
3. To monitor the fluid properties (resistive, capacitant, optical) while pumping out to determine whether oil, water, or gas is entering the chamber
4. As dual packer assemblies run to create a "mini-interference test" that can be used to assess the vertical communication between different intervals.

Pretests and sampling are often not successful. Moreover, the fact that the tool is stationary in the hole for long periods means that there is a higher than usual chance of getting the tool stuck in the hole. One of these problems can occur:

- *Seal failure.* The rubber pad surrounding the probe, which provides a seal between the mud pressure and the formation pressure, may fail, resulting in a rapid pressure buildup to the mud pressure.
- *Supercharging.* Tight sections of the formation may retain some of the pressure they encounter during the drilling pressure (which is higher

than the static mud pressure). The pretest pressure is measured as a pressure that is anomalously high.

- *Dry test.* If the formation is very tight, there may be a very slow buildup of pressure in the pretest chambers, and it is not operationally feasible to attempt to wait until equilibrium is reached.
- *Anomalous gradients.* If sands are isolated even over geological time scales, then they may lie on different pressure trends, not sharing a common aquifer of FWL. Also, if any depletion has occurred in the reservoir or the reservoir is not in a true equilibrium state (for instance, due to a slowly leaking seal or fault), then gradients may not be meaningful.

At this point it will probably be helpful if the distinction is explained among FWL, FOL (free oil level), OWC, GWC (gas/water contact), and GOC and how they are related in pressure measurements (see Figure 2.7.1).

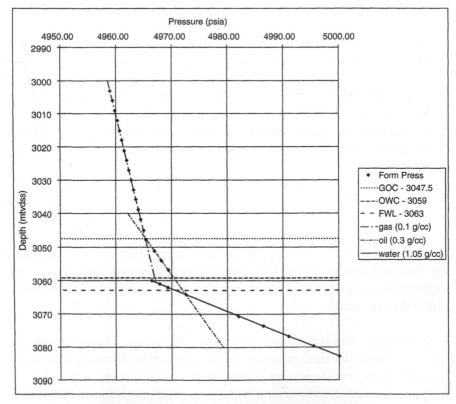

Figure 2.7.1 Example of Formation Pressure Plot

The FWL is the point at which the capillary pressure, P_c, in the reservoir is zero and below which depth no hydrocarbons will be found within that pressure system. Often the FWL may be related to the spill point of the structure, particularly where there is an abundant supply of hydrocarbons in the system. On a formation pressure/depth plot, the intersection between the points of the oil and the water (or gas and water) will fall at the FWL.

Above the FWL, P_c is available to allow the drainage of water by hydrocarbons. However, particularly in low-permeability rocks, a certain entry pressure is required before the value of S_w can fall below unity. Once this pressure is reached, hydrocarbons will be found in the rock and one can be said to be above the OWC or GWC. Note that between the FWL and the OWC/GWC, pressure points will continue to fall on a waterline.

For an oil/gas reservoir, the pressure will rise above the OWC on a trend corresponding to an oil gradient (but intersecting the waterline at the FWL). At the GOC, technically one would expect some kind of similar FWL/OWC effect to occur with an FOL. However, the situation is not the same as at the OWC, because one is dealing with three phases (gas/oil/water) and not two, as before. Hence, it is common practice to treat the GOC as being the same as the intersection point of the gas and oil pressure lines. This may be technically incorrect, but I can only say that it has never caused me any problems during my career as a petrophysicist. For a gas-only reservoir, the pressure will rise above the GWC on a trend corresponding to a gas gradient (but intersecting the waterline at the FWL).

Note that the above considerations have nothing to do with the "transition zone" that relates to the interval between the OWC or GWC and the point at which hydrocarbon values start to approach "irreducible" values. This will be discussed in Chapter 4.

In poor-quality rocks, the effect of entry height can be appreciable (up to tens of meters). It may have the effect of causing the OWC/GWC to vary in depth across the field if the reservoir quality is changing.

2.8 PERMEABILITY DETERMINATION

During a typical pretest, the pressure gauge will show a behavior as shown in Figure 2.8.1.

The behavior of the pressure buildup, analogous to a production-test buildup, may be used to estimate the properties of the formation. The mobility (M) of the formation is defined by:

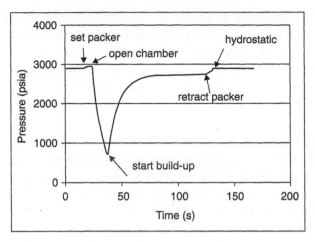

Figure 2.8.1 Pressure Measurements During a Pretest

Table 2.8.1
Typical viscosities of borehole fluids

Fluid	Viscosity (cp)
Water	0.3
Diesel	2–3
Oil	3–10
Gas	0.015

$$M = (k/\mu) \tag{2.8.1}$$

where k = permeability of formation, in md, and μ = viscosity of fluid entering chamber, in centipoises (cp).

It may be shown theoretically that the mobility of the formation is related to the drawdown pressure, drawdown time, and flow rate. From analysis of the buildup, the contractor will normally give a mobility estimate. For conversion of the mobility to permeability, the viscosity needs to be known. In most cases the pretest chamber will be filled with mud filtrate, either water or oil based. Table 2.8.1 gives some values.

While pretests are very useful in that they can prove that some permeability is present if a good buildup is obtained, it should be remembered that they represent only a point measurement. Typically moving the probe up or down by a few centimeters may result in a completely different measurement of mobility. The lack of a good buildup may be purely the result of bad luck in the positioning of the probe. Moreover, the results may not give an accurate idea of the average permeability of a zone.

In general, pretests should be used to verify that a zone has some permeability, but the other methods used (e.g., permeability as derived from a poroperm relationship) are to determine an average permeability to be used in dynamic models. A pretest permeability being lower than that derived from a poroperm relationship may be a result of formation damage occurring while drilling. This may also be observed when the zone is tested for production.

Petrophysicists should always try to obtain the actual field print from the contractor when doing field studies, with a view to assessing permeability and fluid contacts. Reasons for this are as follows:

- Older-generation tools report pressures from a strain gauge, which measures psi per gauge (psig) rather than the absolute psi (psia) reported from quartz gauges. If the values are entered incorrectly into a database, there will be a shift equivalent to atmospheric pressure (14.7 psi).
- When databases are created for fields (e.g., a shared Excel™ spreadsheet), sometimes not all the field data are entered, such as zones reported as "tight." Knowledge of tight zones is crucial if zones are being considered for recompletion based on log-derived permeability estimates.
- When zones are reported as being tight or of limited drawdown, it may be possible in some cases to make an estimate of formation pressure by extrapolating the buildup pressures.
- The contractors will typically report a measured depth for the pretest, as well as a true vertical depth (TVD), with reference to the derrick floor. It is important to check that the pressures used are being referenced properly to the best estimate of TVD relative to the datum (usually mean sea level). After the pressure tool is run, there will typically be a gyro survey run once the final casing is set, and this should be used to convert all measured depths in the well to TVD relative to the datum.

Exercise 2.2. Using Pressure Data

1. Using the formation pressure data points acquired (and presented in Appendix 2), calculate the formation-fluid densities and position of the FWL. Assume that the logs are TVD and measured from mean sea level.
2. Do you interpret the zone to be oil or gas bearing?
3. Given that the sand is laterally extensive, would you propose a production test in this well, and which intervals would you perforate?

FULL INTERPRETATION

The quicklook analyses presented in Chapter 2 will be sufficient for operational decisions on the well. Usually the results are presented by making a clear print of the evaluated logs at scales of 1:200 and 1:500 with the sums and averages marked on the logs and the porefluids marked using appropriate colors. All companies will use blue for water, but some prefer red for oil and green for gas, while others prefer red for gas and green for oil.

Once the final data and prints have been received from the logging contractor, the digital data should be stored within a corporate database. Normally at this point the petrophysicist will do a full interpretation, which might be revised as further core analysis or information from offset wells becomes available.

In some cases the quicklook Archie model might be completely set aside in favor of a more advanced model, as described later in this book. In other cases it is sufficient merely to refine the conventional Archie interpretation. In this chapter the ways the Archie model may be refined will be discussed.

3.1 NET SAND DEFINITION

If core data have been acquired, it is essential that the petrophysicist pay a visit to the core shed at the earliest opportunity to inspect the slabbed core. This will provide a check that there are not anomalous zones that have been wrongly allocated to reservoir or nonreservoir status in the interpretation. Where reservoir can be easily identified, one should make measurements of the core to ascertain the exact net sand footage that can be checked against the calculations made on the logs.

In order to match the net sand footage calculated from the logs with that seen on the core, the shale volume (V_{sh}) cutoff may be varied. Core photographs will be taken under both normal and UV light, which can also assist in the determination of net reservoir. Once the conventional core analyses have been completed, one will have regular measurements of core porosity, grain density, and permeability.

If measurements at overburden conditions have been performed, then the conversion factors to convert porosity and permeability to in-situ conditions should be established. If they are not available, one should assume values based on regional data until special core analyses (SCALs) are completed.

In-situ porosity vs. logarithm of permeability should be plotted, if necessary dividing the data according to facies and/or formation such that a single line can be fitted to the data with reasonable accuracy. This yields the so-called **poroperm relationship**, which is usually of the form (in millidarcies [md], porosity as fraction):

$$k = 10^{\wedge}(k_a + k_b * \phi) \tag{3.1.1}$$

where k = permeability of the reservoir. Typical values of k_a and k_b are -2 and 20, respectively.

Using the V_{sh} cutoff chosen, it should be the case that the net sands should not contain porosities much below a level corresponding to 1 md permeability in oil zones and 0.1 md in gas zones. If this is not true, then it may be necessary to apply an additional porosity cutoff to exclude tight zones, which are not picked up purely by a V_{sh} cutoff.

Where core data are not available, it is sometimes helpful to plot the gamma ray (GR) vs. the density log to help to establish the best point to discriminate net from non-net from the GR log. Typically the plot will show a behavior as shown in Figure 3.1.1.

As shale becomes dispersed in the pore space (increasing GR), the density will rise until the point at which the pore space available for free fluids becomes zero. Beyond this point, the amount of shale may still increase until the formation becomes 100% shale, but the density will change only slightly (depending on variation in density between quartz and shale). The correct cutoff point is therefore the point at which the gradient changes, corresponding to zero effective porosity.

If radioactive minerals are present in the sands, deriving V_{sh} from the GR alone will not be appropriate. In such formations it is recommended to use purely a porosity cutoff. In the case of thinly laminated sands, it is

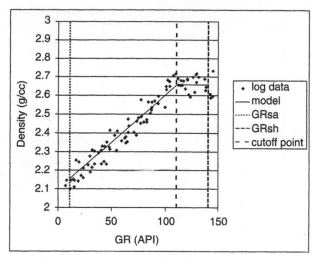

Figure 3.1.1 Determining Reservoir Cutoffs Using a GR-Density Crossplot

possible that the entire formation interval will be designated as nonreservoir using a V_{sh} or porosity cutoff. In this situation it is recommended to not apply any cutoffs whatsoever. The Archie approach will no longer be appropriate, and advanced techniques should be adopted.

3.2 POROSITY CALCULATION

In most cases the density porosity, with an appropriate choice of fluid density, is still recommended. However, a calibration against the conventional core analysis, corrected to in-situ conditions, should be made. The core data should be depth-shifted to match the logs and plotted together with the calculated porosity. A histogram should be made of the core grain density measurements to determine the appropriate value to use in the sands. Note that it is not appropriate to include plugs taken in clearly nonreservoir sections within the analysis. The histogram should provide the mean grain density, as well as give an indication of the likely possible spread of values that could be encountered.

The next step is to make a crossplot of the log density against the in-situ core porosity values, as shown in Figure 3.2.1:

When the core porosity is zero, the density should be equivalent to the core grain density. Also, when the core porosity is unity, the density should be equivalent to the fluid density. The normal procedure is to fix the line

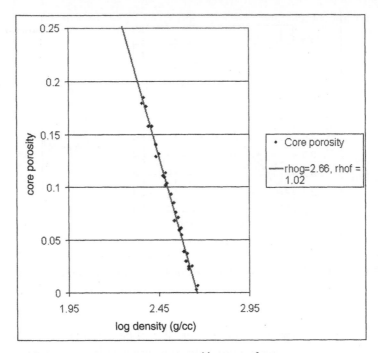

Figure 3.2.1 Core Calibration of Porosity

through the data so that the core grain density is honored, and then to extrapolate the line to the point at which core porosity is unity to determine the appropriate fluid density. Note that this has to be done separately in any gas, oil, and water legs. In theory the fluid densities thus derived should be close to those assumed during the quicklook analyses. Differences might occur due to:

1. Slight miscalibration of the density log
2. Effect of certain mud chemicals (e.g., barite) on the density log
3. Invasion being less or more than previously assumed
4. Problems with the core plug measurements or conversion to in-situ

Whatever the reason for the apparent fluid densities being what they are, the combination of the assumed grain and fluid densities, when applied to the density log, will at least ensure that the log porosities match the core densities. Where the fluid densities are anomalous, one would probably want to use them in only the current well, and possibly only over the cored interval. If, however, the densities agree with the values expected, then they may also be applied with confidence in other wells drilled using similar drilling parameters.

3.3 ARCHIE SATURATION

SCAL data measurements of cementation (m) and saturation (n) exponents should be incorporated into the Archie model. In m measurements, the plugs will have been flushed with a brine of an equivalent salinity to that expected in the reservoir and the resistivity measured. By plotting the logarithm of formation factor, given by $\log(F) = \log(R_o/R_w)$, against $\log$(porosity), according to Archie:

$$\log(F) = -m * \log(\phi) \tag{3.3.1}$$

Therefore, the gradient of the line gives m. Note that the higher the m value used, the higher the water saturations, S_w, that will be calculated, and vica versa.

In n measurements, the plugs will have been flushed with brine, then desaturated (either with air or kerosene) to yield measurements of true resistivity, R_t, vs S_w. By plotting the logarithm of the resistivity index, given by $\log(I) = \log(R_t/R_o)$, against $\log(S_w)$, according to Archie:

$$\log(I) = -n * \log(S_w). \tag{3.3.2}$$

Therefore, the gradient of the line gives n. Note that the higher the n value used, the higher the S_w that will be calculated, and vica versa. Values of n that are anomalously high (above 2.5) may be indicative of a mixed or oil-wet system and require further investigation. Low values of n correspond to good-quality water-wet permeable rock.

Having set m and n, there is no longer complete freedom to choose R_w if one is required to calculate $S_w = 100\%$ in known water sands. If formation-water salinity is well known from produced water samples, one is sometimes faced with a dilemma of whether to honor m or R_w. In many cases, the true cause of this discrepancy is actually an error in the porosity calculation. However, where the porosities are robust, one has to make a choice whether to change m or R_w. It is always worth looking again closely at the cementation-exponent measurements to see how much scatter in the data there is and whether or not the m value chosen is really reliable. If the measurements do not come from the water leg at all, it is possible that diagenetic effects in the reservoir mean that values from the oil leg are not representative. However, it may also be the case that the R_w value is not robust for reasons highlighted in Chapter 2 (Section 2.5).

With respect to the value of R_t to be used, one needs to decide whether invasion or shoulder-bed effects are significantly affecting the deepest reading resistivity tool. For a well drilled with OBM (oil-based mud) that encounters thick sands, I would recommend simply using the deep resistivity as it is. Where there are significant effects of invasion or shoulders, then generally I would always recommend going with a saturation/height approach in favor of Archie.

If it is decided to still try to correct the resistivity for such effects, then the contractor's chart books may be used for making the appropriate corrections, or computer-based algorithms may be applied. Remember that such corrections apply equally in the water leg if one is using a Pickett plot to determine m and R_w.

3.4 PERMEABILITY

For the final evaluation, a permeability log, as well as zonal averages, will usually be required for input to the static and dynamic models. Using the poroperm relationship described in Section 3.1, it is relatively simple to derive a permeability log using the porosity log. However, once the log has been derived, it is important to scrutinize it for any intervals for which the permeability goes to an anomalously high value. Most sandstones do not exceed about 1500 md, although top-quality sands with porosities above 35% may have permeabilities up to about 4000 md. If necessary, apply a cutoff to cap the permeability at a level that is supportable by the core data. In the nonreservoir sections, the permeability should usually be set to a very low value (e.g., 0.001 md). Permeabilities calculated should be roughly in line with those calculated from other sources, such as a formation pressure tool, NMR (nuclear magnetic resonance), or production tests.

For making zonal averages of the permeability, it should be noted that three types of average are possible: arithmetic, geometric, and harmonic. The arithmetic average is given by:

$$k_{arith} = \sum k_i * h_i / \sum h_i \tag{3.4.1}$$

Hence, if, say, the zone were 50 ft, comprising $100\,k$ values at 0.5-ft spacing, the average would simply be the sum of the 100 values divided by 100. This average is appropriate to use if the flow in the reservoir is in the direction of the bedding plane. Small, impermeable streaks will have only very little effect on the average.

The geometric average is given by:

$$k_{geom} = \exp(\Sigma \log(k_i) * h_i / \Sigma h_i) \qquad (3.4.2)$$

In effect, one takes the average of the logarithms of the individual k's, then takes the exponential at the end. This average is appropriate to use if the flow in the reservoir is partially in the direction of the bedding plane and partly normal to it. Impermeable streaks will have some influence but not completely kill off the zonal average.

The harmonic average is given by:

$$k_{harm} = 1/(\Sigma(h_i / k_i) / \Sigma h_i) \qquad (3.4.3)$$

In effect, one takes the average of the inverse of the individual k's, then inverts the result at the end. This average is appropriate to use if the flow in the reservoir is normal to the direction of the bedding plane. Impermeable streaks will completely dominate the zonal average.

Depending on which method is used, the petrophysicist can get widely different results. Typically the arithmetic average will be at least 10 times higher than the harmonic, with the geometric lying somewhere in between.

Note that in horizontal wells there is an additional effect due to the fact that k_v / k_h on the microscopic scale is usually less than 1. The effect of this may be estimated as follows. Let $\alpha = k_v / k_h$, where k_v = permeability of the vertical well and k_h = that of the horizontal. It may be shown that the average permeability (k_{av}) seen by the wellbore, which will be partially influenced by k_v and partly by k_h, is given by:

$$k_{av} = (k_h/2 * \pi) * \int_0^{2\pi} \mathrm{sqrt}(\cos^2(\theta) + \alpha * \sin^2(\theta))d\theta = k_h * (1+\alpha)/2$$

The result, for various values of α, is shown in Figure 3.4.1.

The parameter of k_v / k_h will generally be assumed over an entire reservoir within a dynamic model. Typical values are between 0.1 and 0.3. From Figure 3.4.1, it may be seen that the permeabilities, as determined from a poroperm relationship, need to be adjusted in a horizontal well, even if the formation appears homogeneous throughout.

When giving zonal averages, it is usual to also include the product $k*h$, where h is the thickness of the zone, since it is this which can be

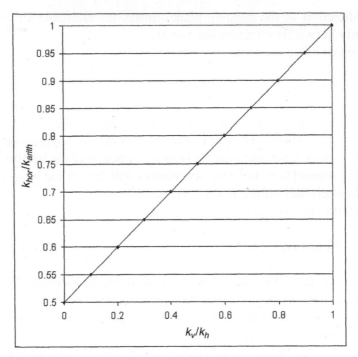

Figure 3.4.1 Effect on Average Permeability of $k_v / k_h < 1$ in horizontal well

related to the flow generated in a production test. Where log-derived values of $k*h$ are compared with production tests, it is often the case that the result derived from the arithmetic average will be higher than that seen in the production test. Reasons for this are:

- Not all the perforated zone contributes to the flow, and the actual h is smaller than that assumed in the petrophysical calculation.
- Some of the flow is not parallel to the bedding plane.
- Formation damage (called "skin") has occurred between the openhole logging and the testing operation. While the test analysis seeks to iden- tify this as a separate term from $k*h$, it may still be partly incorporated into the calculated kh quoted.
- Relative permeability effects, such as gas blocking, may be occurring, making the lab-calibrated in-situ brine permeability inappropriate.

Differences between the log-derived and test permeabilities are a fact of life in real reservoirs and do not necessarily invalidate the log-derived permeabilities that find their way into the static and dynamic models. What happens in practice is that during the history-matching process in the simulator, the permeabilities may be adjusted either globally or near certain wells in order to make the predicted flow rates match the production data.

Exercise 3.1. Full Evaluation of the Test1 Well

Use the data in Appendix 2 to:

1. Revise your net sand discrimination criteria if necessary
2. Calibrate the density log against core porosity. Assume a net effective stress of 2000 psi.
3. Derive a poroperm relationship
4. Derive revised values of m and n to use
5. Recalculate the sums and averages and additionally calculate average permeabilities (using arithmetic, geometric, and harmonic averaging) and $k*h$
6. Calculate the total equivalent hydraulic conductivity (EHC) (thickness $*$ porosity $* S_h$) and compare with your quicklook results.

SATURATION/HEIGHT ANALYSIS

The reason for putting this chapter before other advanced interpretation techniques is that I believe it is of primary importance in correctly defining the STOIIP (stock tank oil initially in place) or GIIP (gas initially in place) of a field. Indeed, because of the way that dynamic reservoir models are constructed in many fields in practice, it completely supersedes any exotic models constructed by the petrophysicist for calculating saturations.

Many times in my career I have seen petrophysical departments working in isolation constructing fabulously complicated models to calculate saturations. But when you ask the geologist what saturations have gone into the static model, he will tell you that he is using a constant value unrelated to the zonal weighted averages. The reservoir engineer may be using one P_c/S_w table in the simulator based on just one air/mercury capillary pressure measurement that he felt was representative of the reservoir in general.

I believe perhaps the most important role the petrophysicist has in a petroleum engineering department concerns ensuring that the saturation/height function being used in the static and dynamic models represent the best possible combination of core and log data, combined with sound petrophysical judgment. In my view, such a function should have both porosity and permeability as input variables, together with height (which may be directly related to P_c).

There are dozens of different functions that have been used to describe capillary behavior in rocks. I have used many of these over my career, but I have found the Leverett J function to be the most broadly applicable. I therefore propose to describe how such a function may be constructed for a reservoir, using both core and log data. This function may be stated thus:

$$S_w = S_{wirr} + a * J^b \tag{4.1}$$

where

$$J = P_c * \left[\sqrt{(k/\phi)} \right] / (\sigma \cos(\theta)) \qquad (4.2)$$

$$P_c = (\text{rho}_w - \text{rho}_h) * h * 3.281 * 0.433 \qquad (4.3)$$

S_{wirr} = irreducible water saturation
rho_w = formation water density, in g/cc
rho_h = hydrocarbon density, in g/cc
P_c = capillary pressure, in psi
k = permeability, in md
ϕ = porosity (as fraction)
σ = interfacial tension between the hydrocarbon and water, in dynes/cm^2
θ = contact angle between the hydrocarbon and water, in degrees
h = height above the free water level (FWL), in m
the constants a and b are to be fitted to the data.

Note that units are not particularly important, as long as they are used consistently throughout. If, say, pressures are used in bars instead of psi, the effect will be for a and b to be modified, but the results will be the same.

4.1 CORE CAPILLARY PRESSURE ANALYSIS

The results of a SCAL program of P_c measurements will usually be presented in the form of Table 4.1.1.

Note that figures in the body of the table represent S_w values. Note also that these measurements will have been performed by one of a number of methods, none of which use actual formation fluids. Use the following steps to generate the average J function. Let:

$$S_{wr} = S_w - S_{wirr}. \qquad (4.1.1)$$

1. Convert the table above into a table of J vs. S_{wr}. Set S_{wirr} equal to 0.01 below the lowest water saturation seen anywhere in the reservoir in cores or logs. In order to derive J, use the values for k and ϕ in the table. For the interfacial tension and contact angle, use the data in Table 4.1.2 depending on the type of measurement used:
2. Plot Log(J) vs. Log(S_{wr}). The intercept and gradient should give you the constants a and b (Figure 4.1.1).

Table 4.1.1
Example of core-derived drainage capillary pressure curves

ϕ	K	P_c (psi)					
		3.000	10.000	25.000	50.000	125.000	200.000
0.078	0.347	0.850	0.783	0.614	0.491	0.386	0.352
0.084	0.992	0.839	0.745	0.525	0.386	0.295	0.269
0.100	2.828	0.763	0.488	0.371	0.281	0.233	0.210
0.096	8.782	0.659	0.353	0.261	0.216	0.201	0.200
0.107	18.350	0.548	0.304	0.218	0.170	0.164	0.165
0.108	11.609	0.651	0.325	0.237	0.198	0.191	0.193
0.123	42.215	0.457	0.270	0.180	0.158	0.155	0.155
0.125	60.976	0.566	0.348	0.258	0.241	0.204	0.200
0.126	157.569	0.377	0.225	0.147	0.127	0.121	0.120

Table 4.1.2
Typical values of interfacial tension and contact angle for lab conditions

	σ (mN/m)	Cos(θ)
Air/mercury	480	0.765
Air/brine	72	1.0
Kerosene/brine	48	0.866
Air/kerosene	24	1.0

Having defined S_{wirr}, a, and b, you basically have all the information you need to construct a saturation/height function for any given porosity and permeability. It is often convenient to create a special poroperm-type relationship using the plug k and ϕ values (Figure 4.1.2). Such a relationship should have the form:

$$k = 10^{(ka+kb*\phi)}. \tag{4.1.2}$$

When constructing J in the reservoir, you will need to use the σ and θ values corresponding to your reservoir conditions. Since these are generally not known, it is recommended to refer to Table 4.1.3.

Having defined such a relationship, it is possible to produce a set of generic saturation/height functions for a range of porosities typically encountered in the reservoir, as shown in Figure 4.1.3:

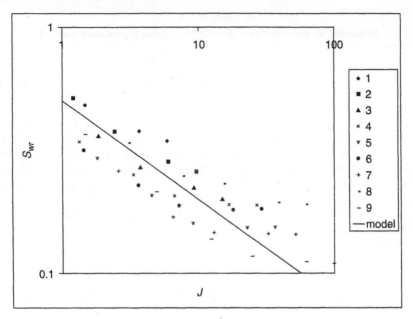

Figure 4.1.1 Fitting a *J* function to Core Data

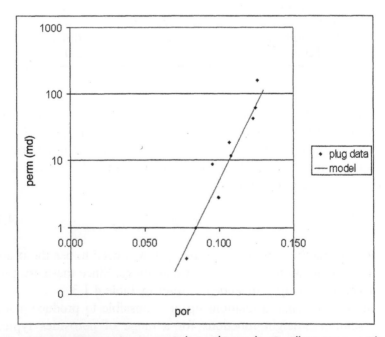

Figure 4.1.2 Fitting a Poroperm Relationship to the Capillary Pressure Plugs

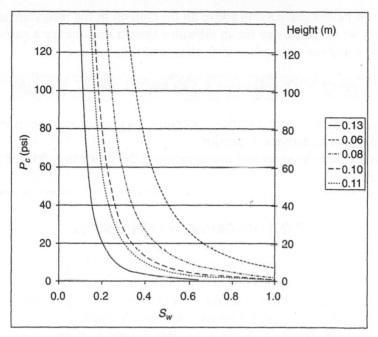

Figure 4.1.3 Generic Saturation/Height Curves from the *J* Function

Table 4.1.3
Typical values of interfacial tension and contact angle for reservoir conditions

	σ (dynes/cm²)	Cos(θ)
Gas/water	50	1.0
Oil/water	30	0.866

These curves, in the form of P_c/S_w tables for particular porosity classes, may be handed directly to the reservoir engineer for inclusion in a simulator. If a geologist wants an "average" value of S_w to use for a static model, the curves should be averaged over a realistic column height for a particular porosity class. Better still is to obtain an area/height table from the geologist and to further weight the curves according to their relative areas at a given height. Since most reservoirs are wider at the base than at the crest, this results in a higher average water saturation.

If you have a gas column above an oil column in the reservoir, generate two set of tables, one for an oil/water system and one for a gas/water system using the appropriate σ, θ, rho_h, and rho_w values.

Exercise 4.1. Core-Derived J Function

1. Based on the core capillary pressure measurements presented in Appendix 2, derive a *J* function.
2. Recalculate the sums and averages using the *J* function derived from these core data.

4.2 LOG-DERIVED FUNCTIONS

The methodology applied to core data can also be applied to the logs to derive an independent *J* function, provided that the position of the FWL is known. This function is useful for comparison with the core-derived function and may be used if no core data are available.

A secondary benefit is that it is possible during the fitting procedure to eliminate thin bed and invasion effects and provide a means of generating a high-resolution saturation log that depends on only the density log.

In order to fit a *J* function, follow these steps:

1. Use the depth of the FWL and the true vertical depth subsea (TVDss) to generate a curve for the height above FWL.
2. Use equation 4.3 to derive a curve for P_c.
3. Use equation 4.2 to derive a curve for *J*. It is assumed that you already have some kind of poroperm relationship that can be used to relate k to the porosity.
4. Use equation 4.4 to derive a curve for S_{wr}. Set S_{wirr} to be 0.01 less than the lowest S_w seen on the logs.
5. Make a crossplot of Log(*J*) vs. Log(S_{wr}).

You will find a cloud of points. Note that shoulder-bed effects and invasion of water-based mud will always pull points to higher S_{wr} values compared with points originating from thick beds and for which less invasion is occurring. Therefore, by fitting the constants a and b so that the model follows the leading edge of the cloud of points, you will effectively be correcting for thin bed and invasion effects.

Having derived a and b, you are now in the same position as you were at the end of the P_c-curve averaging exercise. Generic saturation/height functions can be derived in an identical manner as before. The curves should be compared with those derived from P_c, or "cap," curves and an attempt made to explain any differences. Differences may arise from the following:

- The core measurements, particularly if performed using air/mercury, may be unrepresentative of the reservoir as seen by the logs.
- There may be a discrepancy in the permeability transform used for the core plugs compared with that used on the logs.
- The assumed position of the FWL may be wrong.
- The logs may be influenced by other effects, making either the porosities or the saturations calculated erroneous.
- Depletion may have occurred prior to logging.

In cases where the discrepancies cannot be explained, it may be considered reasonable to take the more pessimistic set of functions as a "low" and the optimistic ones as a "high" and construct a baseline lying halfway between.

Having decided on a baseline function, the logs should be reevaluated using the J function in reverse to recalculate saturations. Considering that the function will use only the curves for height above FWL and porosity/k as input (all derived from the density log), the results may prove surprising. Having applied this technique on more than 20 reservoirs, I have always been amazed at how reasonable the results look, and often how well they compare with the resistivity-derived S_w values, except in the thin beds, where resistivity-derived S_w is usually too high. The density log will typically have a vertical resolution of 1 ft, far superior to the deep induction or laterolog. The resulting increase in average oil or gas saturation will typically be on the order of 20%.

I believe that a properly calibrated Leverett J function, reconciling reliable log saturations with core data, represents the best possible way to calculate saturations in a reservoir and to transfer this information into static and dynamic models. It should also be recognized that where core data are reliable but the position of the FWL is unknown, the function may be used to predict the depth of the FWL from the log saturations. In general I have not had much success in applying this technique, but it may merit further investigation.

Exercise 4.2. Log-Derived *J* Function

1. Use your results from Exercises 2.1 and 3.1 to derive a *J* function based on the log data.
2. Recalculate the sums and averages using the *J* function derived from the logs.
3. Do you recommend using the function derived from the core or log for future STOIIP determination? What are your reasons?

CHAPTER 5

ADVANCED LOG INTERPRETATION TECHNIQUES

5.1 SHALY SAND ANALYSIS

Shales can cause complications for the petrophysicist because they are generally conductive and may therefore mask the high resistance characteristic of hydrocarbons. Clay crystals attract water that is adsorbed onto the surface, as well as cations (e.g., sodium) that are themselves surrounded by hydration water. This gives rise to an excess conductivity compared with rock, in which clay crystals are not present and this space might otherwise be filled with hydrocarbon.

Using Archie's equation in shaly sands results in values of water saturations, S_w, that are too high, and may lead to potentially hydrocarbon bearing zones being missed. Many equations have been proposed in the past for accounting for the excess conductivity resulting from dispersed clays in the formation, which can have the effect of suppressing the resistivity and making S_w calculated using Archie too pessimistic. While these equations will be given, I propose to work only one method through in detail, namely a modification to the Waxman-Smits approach. I have successfully used this method in a number of fields, and it has the advantage of not necessarily relying on additional core analyses for calibration (although these data may be included in the model).

Waxman-Smit's equation may be stated as follows:

$$S_w^{-n^*} = [(R_t/R_w) * \phi^{m^*} * (1 + R_w B Q_v / S_w)] \tag{5.1.1}$$

where B is a constant related to temperature, and Q_v = cation exchange capacity per unit pore volume. Here m^* and n^* have a similar definition as with Archie but are derived in a different way from core data. Temperature determines the excess conductivity (in ohms) resulting for the clay per unit Q_v (in cc).

An equation that may be used to relate B to formation temperature and water resistivity, R_w, as published by Thomas, is as follows:

$$B = (-1.28 + 0.255 * \mathrm{T} - 0.0004059 * T^2)/(1 + (0.04 * \mathrm{T} - 0.27) * R_w^{1.23})$$

$$(5.1.2)$$

where T is measured in degrees Celsius.

Q_v (in meq/unit pore volume in cc) is related to the cation exchange capacity (CEC) (in meq/100 g) of the clay as measured in a laboratory. Q_v may be derived from the CEC using:

$$Q_v = \mathrm{CEC} * \mathrm{density}/(100 * \phi) \qquad (5.1.3)$$

CEC can be measured by chemical titration of crushed core samples. It is dependent on the type of clay. Typical values for clay types are shown in Table 5.1.1.

It has to be said that core-derived measurements based on crushed samples are probably unrepresentative, since the crushing process will expose many more cation exchange types than will be available in the formation. Moreover, I have never been able to relate Q_v to any log-derived parameter (e.g., porosity, V_{sh}) with any success.

A further uncertainty relates to the factor B, which is typically derived using a "standard" correlation that may or not be applicable. A far better approach is to derive the combined factor BQ_v from logs in a known water-bearing sand. I will now present a useful method of doing this. First of all, the assumption is made that BQ_v obeys an equation of the form:

$$BQ_v = (\phi_c - \phi)/(C * \phi). \qquad (5.1.4)$$

Table 5.1.1
Typical Properties of Clays

Clay	CEC (meq/100 g)	Grain Density (g/cc)	Hydrogen Index
Kaolinite	3–15	2.64	0.37
Illite	10–40	2.77	0.09
Montmorillonite	80–150	2.62	0.12
Chlorite	1–30	3.0	0.32

In effect this equation makes the assumption that there is a "clean porosity" (ϕ_c) and that reduction in the measured porosity is as a result of dispersed clay. The excess conductivity BQ_v is assumed to be related to this proportional porosity reduction via the constant C.

If equation 5.1.3 is inserted into equation 5.1.1 and S_w is set to 1 (for water sands), the equation may be rearranged as:

$$C_{wa} = \phi^{-m^*}/R_t = 1/R_w + (\phi_c/C)*(1/\phi - 1/\phi_c) \tag{5.1.5}$$

Hence if C_{wa} is plotted against $1/\phi$ for water sands, the points should fall on a line such that:

- The C_{wa} value at the start of the data cloud represents $1/R_w$.
- The $(1/\phi)$ value at the start of the data cloud represents $1/\phi_c$.
- The points should fall on a gradient equal to (ϕ_c/C).

Note that if there are no shaliness effects, the points should simply create a horizontal flatline from which $1/R_w$ can be read off, with the factor C becoming infinite. If the data fail to fall on a single trend, as above, then the method may be deemed to be inappropriate. An example of some data plotted in this way is shown in Figure 5.1.1.

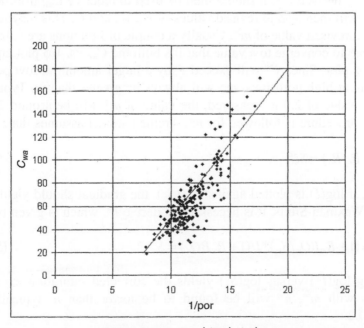

Figure 5.1.1 $C_{wa} / (1/\phi)$ Plot

This plot was made assuming an $m*$ value of 2.0. From the plot it was possible to determine: $R_w = 0.05\,\text{ohmm}$; $\phi_c = 0.13$; and $C = 0.01$. Using an assumed value of $n*$, it is then possible to calculate Waxman-Smits saturations in the hydrocarbon leg.

If SCAL (special core analysis) data are available, it is possible to derive $m*$ and $n*$ from the experiments as follows. In a conventional cementation-exponent (m) measurement, the formation factor F is plotted against ϕ on a log-log scale. In Archie's model, the following would be true:

$$F = (R_o / R_w) = \phi^{-m} \tag{5.1.6}$$

where R_o is the resistivity of the 100% water-saturated rock. Hence $\log(F) = -m*\log(\phi)$ and the gradient of the line yields m. Since for the Waxman-Smits equation it is clearly not the case that $F = \phi^{-m\,*}$, a correction must be made. Let:

$$F* = (1 + R_w B Q_v) * F = \phi^{-m*} \tag{5.1.7}$$

Now, if $F*$ is plotted against ϕ on a log-log scale, it is indeed the case that the gradient yields $m*$.

Having derived $m*$, it should then be used to rederive C_{wa} from R_t and ϕ. This will then lead to revised values of R_w, ϕ_c, and C. This may in turn lead to a revised value of $m*$. Usually a couple of iterations are sufficient to get $m*$ to converge to a value that fits both the C_{wa} vs. $1/\phi$ plot and the $F*$ vs. ϕ plot. Since $F*$ will exceed F by a larger amount at low porosities than at high porosities, $m*$ will always be greater than m. Typically, if an m value of 2.0 is measured, the value of $m*$ will be around 2.2. A similar procedure is followed for $n*$. Archie's model assumes that:

$$I = (R_t / R_o) = S_w^{-n} \tag{5.1.8}$$

Hence, if $\log(I)$ is plotted against $\log(S_w)$, the gradient should yield n.

For Waxman-Smits, it is necessary to derive $I*$, which is given by

$$I* = (1 + R_w B Q_v / S_w) * I / (1 + R_w B Q_v) = S_w^{-n}* . \tag{5.1.9}$$

Plotting $\log(I*)$ versus $\log(S_w)$ yields the corrected saturation exponent $n*$. As with $m*$, $n*$ will be found to be lower than n, typically by about 0.2.

Similar to the Pickett plot made with Archie, the value of the hydro-carbon saturations is not very sensitive to the value of m^*, provided that a water sand is used to calibrate the value of R_w. Note that the value of R_w derived may not correspond to the salinity expected from production tests. This value also includes the effect of the clay-bound water, which may be fresher than the free water and will not flow during production. Hence, it is typically found that R_w appears higher than expected.

The effect of using Waxman-Smits will usually be large only for relatively high values of R_w. This is because the factor $R_w B Q_v$ becomes small compared with unity if R_w is small (saline environments). In this situation the calculated S_w will differ only very slightly from that calculated using Archie's model.

Note that when the equation is applied, a computational complication arises from the fact that S_w appears on both sides of the equation. This can be easily overcome as follows. Initially assume that the value of S_w in the right-hand side of equation 5.1.1 is unity. Calculate S_w and reinsert the new value of S_w into the right-hand side of the equation. Continue in this way until the S_w on the left-hand side ceases to change beyond 0.001 with successive iterations. Typically, five or so iterations are sufficient.

Another way to apply Waxman-Smits method is by the so-called normalized Q_v method, as proposed by Istvan Juhasz. Readers are recommended to read the relevant paper from the Society of Petrophysicists and Well Log Analysts that covers this method in detail (see references). A condensed version will be given here. Juhasz shows that the Waxman-Smits equation may be rearranged in the form:

$$S_w = [(\phi^{-m^*}/R_t)*(S_w * R_{wsh} * R_w)/(Q_{vn} * R_w + (S_w - Q_{vn})* R_{wsh})]^{1/n^*} \quad (5.1.10)$$

where

$$Q_{vn} = V_{sh} * \phi_{sh}/\phi \qquad (5.1.11)$$

$$1/R_{wsh} = \phi^{-m^*}/R_{sh} \qquad (5.1.12)$$

R_{sh} = resistivity of the shale
ϕ_{sh} = porosity of the shale.

The parameter m^* may be determined from plotting $\log(R_t)$ vs. $\log(\phi)$ in water-bearing shaly zones (not clean zones), since the slope of the line is equivalent to m^*. The parameters R_{wsh} and R_w may be determined from plotting C_{wa} vs. Q_{vn}, since the intercept of the points for $Q_{vn} = 0$ on the

C_{wa} axis gives $1/R_w$, and the intercept for $Q_{vn} = 1.0$ gives R_{wsh}. Other equations that are commonly used are:

Dual-water model:

$$1/R_t = S_w^2/(F * R_w) + BQ_v * S_w/F \tag{5.1.13}$$

where

$$F = \phi^{-m}. \tag{5.1.14}$$

Simandoux:

$$S_w^n = A * R_w * \phi^{-m} * \{1/R_t - V_{sh} * S_w/R_{sh}\} \tag{5.1.15}$$

where R_{sh} is the resistivity of the shale.

Indonesia equation:

$$S_w = R_t^{(-1/n)} * \{V_{sh}^{(10V_{sh}/2)}/\sqrt{R_{sh}} + \phi^{(m/2)}/\sqrt{(A * R_w)}\}^{(-2/n)} \tag{5.1.16}$$

Since Waxman-Smits fulfills all the criteria I require from a shaly sand equation (i.e., it introduces a clay conductivity element that is related to the amount of clay as determined by the logs), I have only rarely used other equations during my career. As stated in Chapter 3, I would always prefer to derive saturations for STOIIP (stock tank oil initially in place) and GIIP (gas initially in place) using a saturation/height function calibrated against good-quality core measurements.

Exercise 5.1. Shaly Sand Analysis

1. Using data from the water leg in the test1 well, make a relationship of BQ_v to porosity. Also derive R_w from the plot.
2. Using the core data in Appendix 2, calculate m^* and n^*.
3. Calculate saturations using the Waxman-Smits equation and make a new table of sums and averages.
4. How do the Waxman-Smits saturations compare with those derived using Archie and those from the core-derived J function?

5.2 CARBONATES

Most of the analyses presented in earlier chapters work equally well with carbonate formations. However, the grain density used for deriving porosity from the density log needs to be set to a different value. For limestones the most common value used is 2.71 g/cc, although a direct measurement on the core is preferred, if available. Dolomite and anhydrite have higher matrix densities (see Appendix 6). In many fields with good-quality carbonate reservoirs, evaluation is simpler than with sandstone because the shaly sand problem does not occur and beds tend to be thicker. In general, it will be found that Archie's model works very well.

However, during my career I have also worked on some extremely tight gas- and oil-bearing carbonate reservoirs where normal techniques did not give good results. I will therefore discuss these. Typical characteristics of low-permeability (<1 md) carbonate reservoirs are:

- Porosity may still be appreciable but be in the form of isolated pockets or vugs. This is often seen in chalk reservoirs.
- Entry heights and transition zones may be extremely long (up to 100 m).
- Matrix permeability may be extremely low, but the well may flow due to the presence of natural (or mechanically induced) fractures in the formation. Due to the fact that these fractures may cause significant losses during drilling, the well may need to produce back a lot of drilling fluid before formation hydrocarbons start to flow.

In such a field, the development strategy will often be to drill horizontal wells perpendicular to the natural fracture orientation, thereby optimizing the productivity of the wells. The STOIIP or GIIP will be stored mainly in the matrix but flow in the well via the fractures. Hence, if there is not an extensive fracture system that provides a conduit for the matrix to flow into, the wells will either water out or die very quickly.

Clearly the ability to detect the presence and orientation of fractures is extremely important. During drilling it may be useful to monitor the extent of losses, and depths at which they start to occur, which may provide valuable information as to the presence of fractures. Needless to say, the measures adopted by drilling to cure these losses need to be examined to ensure that any fracture permeability is not permanently impaired. Conventional logs that may give indication of fractures are:

- Sonic log (cycle skipping occurrence)
- Microresistivity tool (erratic behavior as the pad crosses a fracture)

- GR (spikes occurring where fractures have become cemented up with radioactive minerals)
- Caliper (borehole will tend to become elliptical, with the major axis perpendicular to the tectonic stress direction)

In order to properly characterize the types and orientation of fractures, it is necessary to use imaging tools. Resistivity tools are preferred because it is easier to differentiate open fractures filled with fluid from cemented fractures. However, ultrasonic-based tools can also be used and are the only option in oil-based mud (OBM).

Note that naturally occurring fractures will tend to be oriented in the direction of maximum horizontal stress in the field. Particularly in areas close to major fault systems, the difference between the stresses in different directions may be very large and conducive to fracturing. Cores may also prove invaluable in characterizing any fractures that may be present, although care has to be taken to exclude drilling-induced fractures. It is also necessary that an orientation tool be run with the coring assembly. Once the fracture system has been analyzed, it is useful to derive a fracture density curve that may be included with the other logs. Such a curve may be correlated with the horizontal permeability k_h derived during a well test and used to predict the producibility of future well penetrations.

Carbonate reservoirs, unlike clastic reservoirs, may well be amenable to HCl acidization treatments either with or without mechanical fracturing ("fraccing") of the reservoir. However, such treatments can affect only the region around the wellbore and will not compensate for poor permeability and lack of fractures over the wider extent of the field. Pressure testing/sampling of tight carbonate reservoirs using a conventional probe is nearly always unsuccessful. In order to have any hope of success, one would need to use a packer-type of tool, and even then the success rate is typically low.

In chalk reservoirs, compaction may well be an issue during field life. While this has the advantage of providing an additional pressure support mechanism, extensive studies will be required during the design phase of any installations, particularly offshore.

5.3 MULTIMINERAL/STATISTICAL MODELS

As stated in Chapter 2, my preferred method of calculating porosity is the density log using the appropriate matrix and fluid densities. This approach can go badly wrong if heavier minerals are also present in the formation in variable amounts. I have worked in some fields having varying amounts of limestone, marl, anhydrite, dolomite, siderite, pyrite, quartz, and clays where a conventional approach using deterministic equations is not reliable. In such a situation, the best approach is to adopt a multimineral/statistical model. The basic way programs using this technique operate is as follows:

1. The various minerals and fluids to be included in the model are determined.
2. The response of each of these minerals/fluids for a variety of parameters as measured by logging tools is specified by the user.
3. The program finds the combination of mineral/fluid volume fractions that most closely matches the observed log responses, such as to a variety of criteria and constraints specified by the user, such as the:
   - relative importance (weighting) of various tools,
   - measurement error for each tool,
   - relative saturations of fluids in the invaded zone as opposed to the virgin zone,
   - relative amounts of various minerals, and
   - resistivity response relative to fluid saturations (e.g., Archie, dual-water, Waxman-Smits, etc.).

Needless to say, it is not possible to have a greater number of minerals/fluids than the number of tool response equations (although the fact that the sum of all the volume fractions must equal 1 effectively provides an extra equation).

Overall, this sounds like a very rigorous approach to a conventional sort of deterministic evaluation, which would be preferred in all cases. Reasons why it is not are as follows:

1. The fact that the program is capable of calculating back the correct log responses does not mean that the results are necessarily correct. If a bad choice of minerals/fluids is made or their properties are incorrect, a solution may be found that is completely wrong.
2. The program does not have any depth-dependent reasoning capability. Hence, one may frequently find "a bit of everything everywhere," with gas below oil and oil below water and minerals popping up all over the place, often more in response to the hole quality than anything else.
3. Many of the tool responses for minerals in the formation are not known accurately and recourse is made to standard tables of typical minerals.

It is found that changing the value of one parameter can have a drastic effect on the resulting volume fraction of that mineral.

4. Porosities derived by the program will typically come about from a combination of the density, neutron, and sonic responses, plus whatever assumption is made about the relative saturations in the invaded zone as opposed to the virgin zone. I am dubious about how correct such a porosity really is.

5. Unlike with a deterministic approach, where the resulting uncertainty in porosity and saturation may be directly traced to uncertainty in the input parameters, statistical programs represent a "black box" approach, in which there is no clear audit trail between the input and the output.

6. The programs are very sensitive to log quality and noise on the log traces. Where even one log is reading wrong, the volume fractions will be affected and the output may be completely unreliable.

Overall, my impression is that there are some cases in which statistical models offer real advantages over conventional interpretations. A good example is in a sandstone reservoir having variable amounts of siderite or pyrite. The program is also useful in a normal sandstone reservoir where there are limestone stringers intermittently present. In situations where the mineralogy is not well known or the logs are of poor quality, I am extremely dubious about the quantitative correctness of the output. Even when neither of these situations arises, it is still my experience that it is necessary to make dozens of runs, investigating the effects of minor changes to the input parameters, before a solution can be produced in which one can have any confidence.

All the main logging contractors are able to offer software for statistical analyses, which can be run either by the contractor or in-house by the oil company. The values for typical parameters for various minerals are usually built into the software as default values, so will not be repeated here.

5.4 NUCLEAR MAGNETIC RESONANCE LOGGING

When first introduced in the late 1980s, NMR logging attracted a lot of interest because it was a whole new type of measurement and offered the possibility of direct measurement of porosity and the differentiation of fluid types and the relative contributions arising from clay-bound water from free water. In this respect it offered to solve one of the problems occasionally confronting petrophysicists, namely, low-resistivity pay.

The basic principle by which the tool operates is as follows. The tool is assumed to respond only to hydrogen nuclei (in water, oil, and gas) in the porespace. The hydrogen nuclei (which are just protons) in the pore-fluids have a spin and magnetic moment that may be affected by an external magnetic field. In the absence of an atomic field, these moments are aligned randomly. When an external field (B_0) is applied, a process occurs whereby the orientation of the nuclei changes so that a proportion of them align in the direction of the applied field H. The reason they do not all immediately align in the direction of the field is that two adjacent nuclei are in a lower energy state when they are aligned in opposite directions.

The nuclei do not immediately align in the direction of H, but their spins precess around B_0 at a frequency given by Larmor:

$$\omega_{Larmor} = \gamma B_0$$

where γ is the gyromagnetic ratio (42.58 Mhz/T for hydrogen) and B_0 is the strength of the external field, in Tesla.

After a time a proportion of the nuclei have "relaxed" to be aligned with B_0. The resulting magnetization of the formation is given by M_v, and will vary according to:

$$M_v \propto (1 - \exp(-t/T_1))$$

where T_1 is the longitudinal relaxation time. In the absence of any further fields being applied to the horizontal magnetic components, the individual nuclei will be randomly distributed and sum to zero.

Now consider what happens if, after a period denoted by T_w (wait time), a horizontal magnetic field is applied at a frequency equal to the Larmor frequency. The nuclei's horizontal magnetic moments will start to align themselves in the direction of the horizontal pulse. After a time given by τ (the echo time), a pulse is given at 180 degrees to the direction of B_0 (called a π pulse). The nuclei start to align themselves in the opposite direction. However, because of differences in their horizontal relaxation times, the magnetic moment building up in the opposite direction will be less than during the first pulse. A third pulse is then applied in the original direction of B_0 with correspondingly even less buildup of moment. The process is continued for a finite number of "echoes" until the horizontal signal (which can be detected in the tool's coils) dies away to zero. The decay of the horizontal signal is called the **transverse relaxation**, and the magnetization detected (denoted by M_h, ignoring diffusion) will vary according to:

$$M_h = M_0 \cdot \exp(-t/T_2)$$

This pulse scheme is referred to as a CPMG (Carr, Purcell, Meiboom, and Gill) excitation. In practice, not all the fluid in the pores will relax according to the same T_2. Those lying close to the pore wall will relax more quickly than those in the center of the pore. This means that there is a series of contributions to M_h, each decaying at a different rate. In addition, because of diffusion of the nuclei within the pores, nuclei that may not initially be close to the wall may move toward the wall during the measurement and relax more quickly. This introduces an extra term into the behavior of M_h, given by:

$$\text{Exp}(-t \cdot \gamma^2 \cdot D \cdot G^2 \cdot \tau^2 /3)$$

where D = molecular self-diffusion coefficient and G = gradient of the static magnetic field. Hence the full expression for M_h is given by:

$$M_h = M_0 \cdot \exp(-t/T_2) \cdot \text{Exp}(-t \cdot \gamma^2 \cdot D \cdot G^2 \cdot \tau^2 /3).$$

Because different fluids (oil, gas, water) have different values of D, if measurements are done at different values of τ, there is the possibility of differentiating fluid type. This is the basis for what is called time domain analysis (TDA). The influence of the pore wall on T_2 is assumed to follow a relation of the form:

$$1/T_2 = 1/T_{2,bulk} + \rho \cdot S/V$$

where $T_{2,bulk}$ = relaxation time of bulk fluid, ρ = surface relaxivity, and S/V = pore surface-to-volume ratio. For spherical pores of radius r, S/V reduces to $\rho \cdot 3/r$.

Note that the expression for capillary pressure, P_c, is given by:

$$P_c = 2 \cdot (s \cdot \cos(\theta))/r.$$

Hence it can be shown that if T_2, bulk $\gg \rho \cdot S/V$, which is the case close to the pore wall, then:

$$P_c \cdot T_2 = (\sigma \cdot \cos(\theta)/1.5 \cdot \rho).$$

This is important because it shows how a cutoff between bound and free fluid, made on the basis of P_c, can be translated into a cutoff based on T_2 for distinguishing between bound and free fluid. This is the basis for the T_2 cutoff commonly used: 33 ms for water-wet sandstones and 100 ms for carbonates. The 33 ms is based on an assumed surface relaxivity of sandstones of 100 μm/s. In fact the relaxivity may vary considerably among different types of rock. Values as low as 14.4 have been reported in the literature. Times as long as 200 ms have also been seen in sandstones drilled with OBM.

If the appropriate T_2 cutoff is not a constant but is facies dependent, significant problems are caused in determining permeability accurately. In fact, a conventional poroperm approach normally works quite well if a different relationship is used according to facies type. A lot of the potential benefits from NMR are removed if one requires both core T_2 measurements for all facies types and a means for determining which facies is being logged. The permeability may also be severely affected if, based on TDA, a gas/oil/water model is being assumed rather than a straight oil/water model. Applying the gas correction may affect the permeabilities by a factor of up to 100.

The total porosity of the sample is related to the strength of the initial signal occurring from the tool following the first transverse pulse during a T_2 acquisition. Note, however, that for the following reasons this might read low:

- If the wait time T_w (also sometimes denoted as T_r, the recovery time) prior to the CPMG excitation is too short, the transverse field will be reduced. This is referred to as incomplete polarization, to which polarization correction may be applied.
- The tool is calibrated assuming 100% freshwater in the pores, i.e., the hydrogen index (HI) is 1.0. The HI is influenced by temperature, pressure, and salinity, as well as the fluid type (water, oil, or gas).

Because clay-bound water relaxes very fast (T_2 of a few ms), a special mode of acquisition is required to measure total porosity. In a normal acquisition mode, the tool will respond to only capillary-bound and free fluids. It should be noted that it is normally assumed that the rock is water wet. This means that any short T_2 arrivals are the result of the wetting phase relaxing close to the pore wall. Other fluids, such as gas and water, are assumed to be far from the pore wall, so that one sees only their bulk fluid relaxation times. Any kind of TDA, which exploits the differences in D, exploits this fact. The interpretation of the tool results can be subject to serious errors if this assumption is not true, which can be the case in a well drilled with OBM for which surfactants in the OBM filtrate have made the invaded zone oil wet. Even where WBM (water-based mud) has been used, any mixed wettability in the formation will tend to result in anomalous results.

The tool will measure a decaying magnetic amplitude vs. time, which depends on the following parameters:

- B_0 (the static field strength)
- T_w (or T_r) (the wait time for longitudinal polarization)
- τ (the transverse echo time)
- T_1 of the fluids in the pore space
- T_2 of the fluids in the pore space
- D of the fluids in the pore space
- The total porosity

Note that the signal will arise from only the part of the formation for which the CPMG pulses correspond to the correct Larmor frequency. Because the fixed magnet is located in the borehole, with the magnetic field decreasing with distance from the borehole, this will define the zone of investigation of the tool.

Having measured the transverse signal as a function of time, the next step is to invert these data into the corresponding distribution of T_2 values that make up the signal. This would be a straightforward mathematical operation were it not for the presence of noise in the signal. In fact, without some additional form of constraint, at the noise levels typically encountered in the tool it is possible to produce wildly different T_2 distributions that can all honor the original decay curve. One constraint that is commonly applied, called regularization, is that the T_2 distribution must be smooth. This results in a more stable solution, although there is no particular reason why the T_2 spectrum should indeed be smooth. Needless to say, unless the inversion is correct, the results of the tool will be completely useless. This is worth bearing in mind in situations where the tool gives results that cannot be explained in terms of known properties of the lithology based on core data.

In practice the T_2 spectra are not continuous but divided into "bins" covering different ranges of T_2. The maximum value of T_2 that can be measured is determined by the time allowed for the signal to be measured. This in turn is related to the logging speed. In some situations, wait times of up to 15 seconds might be needed to capture the full spectra, translating into logging speeds that are very slow (under 100 ft/hour).

Output curves common to NMR tools include the following:

T_{por} = total porosity obtained from summing all the fluid contributions
POR_{eff} = effective porosity obtained by excluding contributions prior to a predefined T_2 threshold
BVI = the proportion of capillary-bound water to the total volume
CBW = the proportion of the clay-bound water to the total volume
FFI = free fluid index. The proportion of non-clay-bound or non-capillary-bound fluid to the total volume. POR_{eff} = (BVI + FFI)/100.
PERM = permeability obtained by applying various standard equations

The most commonly used equation for permeability (Coates equation) is as follows:

$$PERM = \left((T_{por}/C)^4 \right) * (FFI/BVI)^2$$

where C is a constant (typical values are around 0.10).

By definition BVI + CBW + FFI = T_{por}. Above the transition zone, i.e., at a P_c value of >100 psi, it should be true that:

$$S_h(\text{hydrocarbon saturation}) = FFI/T_{por}.$$

Below this P_c value, FFI will comprise both the free water and the free hydrocarbon. An example of an NMR log, from a hypothetical formation lying well above the FWL (free water level), is shown in Figure 5.4.1.

Fluid differentiation with the tool may be performed by either the differential spectrum method (DSM) or the shifted spectrum method (SSM). DSM works by varying the wait time T_w, thereby exploiting the different T_1 times for different fluid types. SSM works by varying the echo time T_e, thereby exploiting differences in the diffusivity (D) between different fluids. This is particularly applicable to gas/oil differentiation, since the value of D for gas is much higher than that for oil or water. Typical acquisition parameters for the tool are as follows:

Normal T_2 mode:
- Number of echoes: 600
- Wait time: 1.3 seconds (or longer if a significant polarization correction is required)
- Echo spacing: 0.32 ms
- Logging speed: 600 ft/hr (i.e., 2 measurements at each depth increment)

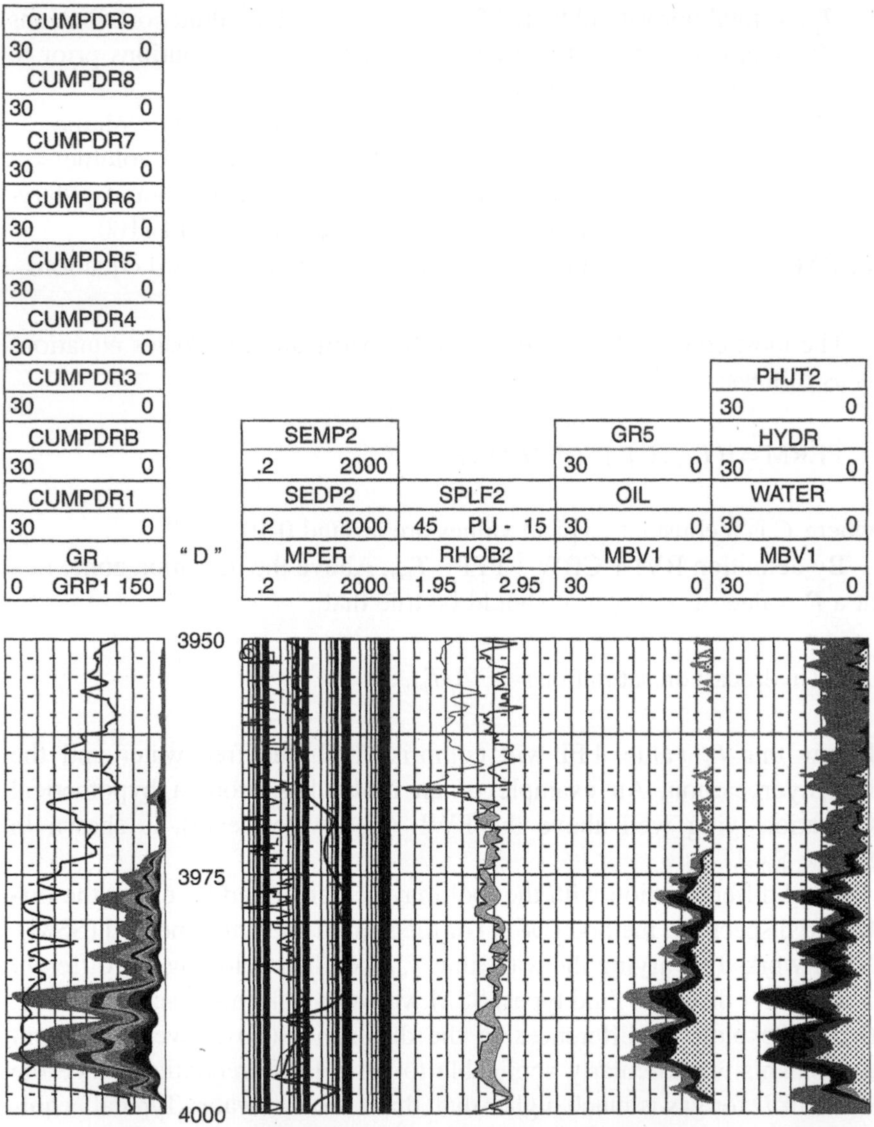

CUMPDR9					
30 0					
CUMPDR8					
30 0					
CUMPDR7					
30 0					
CUMPDR6					
30 0					
CUMPDR5					
30 0					
CUMPDR4					
30 0					
CUMPDR3					PHJT2
30 0					30 0
CUMPDRB		SEMP2		GR5	HYDR
30 0		.2 2000		30 0	30 0
CUMPDR1		SEDP2	SPLF2	OIL	WATER
30 0		.2 2000	45 PU - 15	30 0	30 0
GR	" D "	MPER	RHOB2	MBV1	MBV1
0 GRP1 150		.2 2000	1.95 2.95	30 0	30 0

Figure 5.4.1 Example of an NMR Log

DSM mode:

- Two wait times are chosen in order to detect a component with a long T_1, such as gas or light oil.

SSM mode:

- Two T_e times are used to detect a component with a different D value, such as gas.

Bound water mode:
- Number of echoes: 100
- Wait time: 0.2 ms
- Echo spacing: 0.32 ms

Total porosity mode:
- 50 measurements of 10 echoes are made
- Echo spacing: 0.32 ms
- Wait time: 5 ms

Note that data can also be acquired in a stationary mode. Often a few stationary measurements, with very long wait times and numbers of echoes, are acquired as a check that the logging speed being used is not too fast. NMR properties of reservoir fluids vary with pressure, temperature, salinity, and viscosity. Table 5.4.1 gives some general values that may be of use.

Through further TDAs, the software can produce the oil and gas saturation. If resistivity data are input into the software, a secondary measurement of hydrocarbon saturation is made. Limitations in the physics of early-generation tool were:

1. The difficulty of generating a uniform magnetic field over the parts of the formation from which the measurements were being made (the T_1 and T_2 times are dependent on the strength of the static field)
2. Problems with the static magnetic field magnets at downhole temperatures
3. Problems with logging speed when the relaxation times could be as long as many seconds

Table 5.4.1
Typical NMR Properties of Reservoir Fluids

Fluid	T_1 (seconds)	T_2 (seconds)	D $(10^{-9}\,m^2/s)$
Brine (100 kppm) at downhole conditions (200 bar, 125°C)	20–25	20–25	18–22
Natural gas at downhole conditions	2–3	—	170–180
Oil, viscosity 10 cp	0.1	0.1	0.1

4. The difficulty of obtaining sufficient depth of investigation for the measurements to be useful for saturation determination.

Most of the early drawbacks with the tool have been overcome. However, it still suffers from the following limitations:

- The tool response is still severely affected by the presence of any diamagnetic or paramagnetic ions such as arise from any iron in the mud or formation, or manganese/vanadium. These will have a great effect on the relaxation times of the hydrogen nuclei.
- The tool is typically far more expensive to run than a conventional logging suite, which will normally be run in addition to it anyway, due to lack of confidence in the tool and the need to keep consistent with earlier logs in the field.
- Logging speed is typically 800 ft/hr, less than half that of conventional logging (1800 ft/hr). Hence, there is additional rig time, adding to the total cost.
- There are still temperature limitations and a lack of slimhole tools available. An LWD (logging while drilling) version of the tool is still in the test phase.
- The depth of investigation is still far shallower than the deep-reading resistivity tools.
- Early claims asserted that the tool offered a superior measurement of formation permeability; however, in practice the permeability was derived from an empirical equation involving the FFI and the T_{por}. In most cases the "global" calibration parameters were found to be inapplicable, requiring a local calibration in each formation against core data.
- It has not been conclusively demonstrated that the tool offers a more cost-effective or accurate evaluation in standard simple reservoirs where conventional techniques and models are being applied.

Where it is hoped that the tool may offer a direct advantage over traditional techniques lies in the following areas:

- Identification of zones previously missed due to high percentage of clay-bound water that would nevertheless flow dry hydrocarbons
- A more accurate determination of porosity, particularly in complex lithologies
- Advanced facies discrimination in formations where conventional logs are not capable of discrimination

- A better measurement of permeability than currently possible using traditional poroperm-type plots
- In-situ measurement of oil viscosity
- Differentiation of oil/gas zones
- The elimination of the need to run nuclear sources in the hole

Overall, it may be said that some petrophysicists really believe in the future of NMR logging and see such a tool eventually replacing conventional logs. Others point to the fact that NMR logging has been around for 15 years and has offered few real advantages in most fields. I have seen many NMR logs in which the tool shows oil in known water legs and both gas and oil both above and below the GOC (gas/oil contact). I have also seen permeabilities differ by a factor of 10 or more when compared with core-calibrated values derived from poroperm relationships. However, I have also seen the tool explain why some zones, with high total water saturation, are capable of producing dry oil. Therefore, there are situations in which NMR can offer real advantages, but running the tool should be justified on a case-by-case basis, and not just from a need to be perceived as "high tech."

5.5 FUZZY LOGIC

"Fuzzy logic" is a technique that assists in facies discrimination, and that may have particular application in tying together petrophysical and seismic data. In this chapter, the basic technique will be explained, together with a worked example to illustrate the principle. Consider a situation in which one is using a GR (gamma ray) log to discriminate sand and shale. With the conventional approach, one would determine a cutoff value below which the lithology should be set to sand and above which it should be set to shale. To use fuzzy logic, one would do the following:

1. In some section of the well where sand and shale can be identified with complete confidence, one would generate a "learning set," that is, create a new log in which the values are set to 0 or 1 depending on whether the formation is sand or shale.
2. Over the interval defined by the learning set, one would separate all the bits of GR log corresponding to sand and shale, respectively.
3. For the sand facies, a histogram would be made of all the individual GR readings. To this distribution would be fitted a mathematical function (most commonly a normal distribution) that would capture the

mean and spread of the data points. This is often called a "membership function."

4. The same would be done for all the shale values, generating a new membership function with its own mean and spread.
5. Both membership functions would be normalized so that the area underneath them is unity.

The resulting distributions would look like Figure 5.5.1.

Now, supposing one were trying to determine whether a new interval of formation, having a GR reading of x, belonged to the class "sand" or "shale." Using the functions shown in Figure 5.5.1, one would simply enter the graph at the GR axis at the value x and read off the relative probabilities of the interval belonging to either class. The interval would be assigned to the one having the greatest probability. Moreover, one can assign a confidence level based on the relative probabilities.

Having understood the principle of fuzzy logic with one variable, it is easy to see how it might be extended to more than two classes (e.g., sand, silt, and shale) and with more than one input variable (e.g., GR, density, neutron). Since it is not practical to plot more than two variables on a graph, the actual allocation is performed in a computer program in the N dimensional space corresponding to the N variables. Obviously for the method to work well, it is necessary that the membership function not

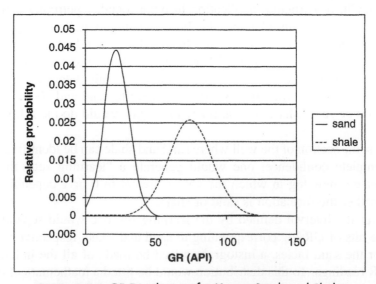

Figure 5.5.1 GR Distributions for Known Sands and Shales

overlap much in the N dimensional space they occupy. Also, the method does not work well with parameters that vary gradually with depth.

The advantage of the method over other approaches such as neural networks is that one is able to see, through plotting the membership functions with respect to a certain variable, whether or not it is applicable to include a certain variable or not. Also, the method can generate a confidence level for the output classification, as well as a "second choice." The use of fuzzy logic has been mainly in acoustic and elastic impedance modeling, where one can investigate whether or not, for instance, there is any acoustic impedance contrast between oil- and gas-filled sandstones. If there is, the membership functions may be used as input to a seismic cube for allocating facies types to parts of the seismic volume, thereby showing up potential hydrocarbon zones.

Fuzzy logic may also be useful to allocate certain facies types to the logs, as for instance a basis of applying a different poroperm model. In my experience with using fuzzy logic, I have often found that one starts out with too many facies, which then are found to overlap each other. Also, the effect of adding more log types as variables, which may be only loosely related to the properties one is interested in, is generally detrimental. In many respects, fuzzy logic is similar to the statistical analyses packages described earlier. In common with these, it has the advantage over deterministic techniques in that it can handle a lot of variables impartially and simultaneously. However, also in common with those packages, it can easily generate rubbish unless great care is taken with the input.

Exercise 5.2. Fuzzy Logic

1. Set up a fuzzy logic model to distinguish between net and non-net on the basis of GR using the data from the core as a learning set.
2. Apply the model to the lower half of the entire logged interval. Compare the average net/gross with that derived using the conventional analyses.

5.6 THIN BEDS

Conventional petrophysics relies on the logs being able to resolve the individual beds in order to determine such properties as R_t, rho$_b$, and GR. While examination of any core will reveal many features that are far below the resolution of all but imaging logs, this commonly does not pose any serious problems, provided that any variations average over an interval used for evaluation.

Problems arise where the variations initiate a nonlinear response in the tool used to evaluate them. The most common example of this is the effect of thin shale beds on the resistivity log. For beds that are perpendicular to the borehole, the resistivity may be approximated by:

$$1/R_t = 1/R_{shale} + 1/R_{sand}$$

Hence a small amount of conductive shale may significantly lower the R_t seen by the tool. When saturations are calculated, they may make the zone appear predominantly water bearing and not perforated, when in fact the water saturations in the individual sand beds are very low and the zone is capable of producing dry oil.

Note that if the laminae are sufficiently wide, so that they are still resolved by the density log (~1-ft resolution), the problem can easily be overcome by using a J function approach, as described in Chapter 3. If the laminae are on the millimeter scale, then they will not be resolved by the density log and other approaches should be adopted.

First of all it is necessary to identify the laminated zones and to determine the proportion of sand to shale. The most reliable way to identify laminated sand is through direct inspection of the core. Measurements should be made to determine the relative thicknesses of the sand and shale layers as a function of depth. If this is done, then it is recommended to assume common properties for the sand, with the porosity taken from a core average and the saturation derived using a saturation/height function from core capillary pressure measurements. If core is not available, then common ways to identify laminated sands are:

- Use of borehole images derived from either resistivity or ultrasonic-based tools
- Inspection of the microresistivity, which may show rapidly varying behavior (although this may also be due simply to variations in the borehole wall)
- Measurement of resistivity anisotropy, either from comparison of induction with laterolog types of devices or by running special induction tools with perpendicular coils, which may be indicative of laminated sequences
- The presence of strong shows while drilling in zones appearing to be nonreservoir on the logs

Note that the situation of the laminae being perpendicular to the borehole represents the worst case in terms of the suppression effect on the resistivity. Where the borehole is inclined to the bedding, the resistivity is affected less, and there are published equations for determining the resistivity measured as a function of the orientation angle and R_{shale}, R_{sand}.

Another technique that has been applied is the Thomas-Stieber plot. This will now be explained. Start by considering a clean sand with porosity ϕ_i and containing water with a hydrogen index of HI_w. There are three ways in which shale can be introduced (Figure 5.6.1):

1. Laminae of pure shale may be introduced in the proportion V_{lam}, with V_{lam} increasing until the point at which the formation becomes 100% shale.
2. Dispersed shale may fill the existing porespace until the point at which the pores are completely filled with shale. The total volume percentage of shale is given by V_{sh}.
3. Structural shale may replace the quartz grains while the primary porosity remains the same. The total volume percentage of shale is given by V_{sh}.

The way these processes will affect ϕ_d and ϕ_n may be predicted as follows. Let the shale porosity be denoted by ϕ_{sh} and the clean sand porosity be denoted by ϕ_{csa}. The HI of the shale is denoted by HI_{sh}. Assume that

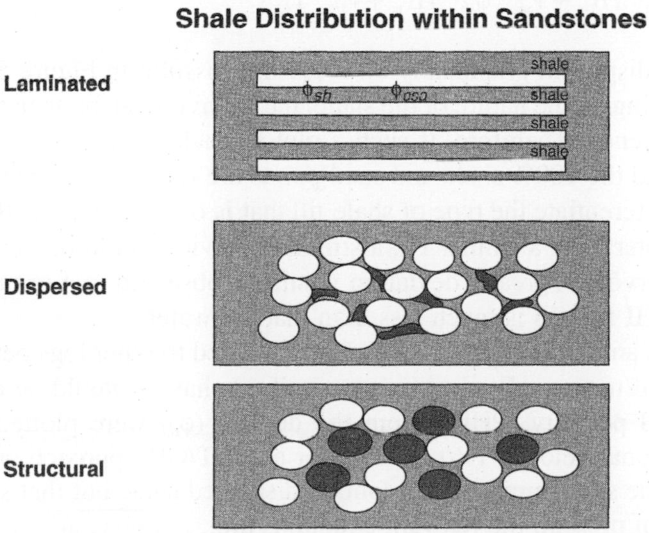

Shale Distribution within Sandstones

Figure 5.6.1 Distribution of Shales Within Sandstones

the shale and quartz have a similar matrix density and that the formation is water bearing.

Laminae:

$$\phi_d = V_{lam} * \phi_{sh} + (1 - V_{lam}) * \phi_{csa}$$

$$\phi_n = V_{lam} * (HI_{sh} + \phi_{sh} * HI_w) + (1 - V_{lam}) * \phi_{csa} * HI_w$$

where V_{lam} is the volume fraction of laminated shale.

Dispersed:

$$\phi_d = \phi_{csa} - V_{sh} * (1 - \phi_{sh})$$

$$\phi_n = HI_{sh} * V_{sh} + HI_w * \phi_d$$

where V_{sh} is the volume fraction of shale.

Structural:

$$\phi_d = \phi_{csa} + V_{sh} * \phi_{sh}$$

$$\phi_n = \phi_{csa} * HI_w + V_{sh} * \phi_{sh} * HI_w + V_{sh} * HI_{sh}$$

This is displayed graphically on a ϕ_d / ϕ_n crossplot in Figure 5.6.2.

Depending on the nature of the shale, the behavior can be seen to follow different trends. Therefore, if such a plot is made over a section of formation, and the 100% shale and sand points are identified, it may be possible to differentiate the type of shale fill that is occurring. Note that since oil and water have a similar HI, a similar behavior would be observed in an oil reservoir. A greater deviation would be observed in a gas reservoir since the HI for gas is much less than that for water.

Thomas and Stieber's method can be extended to other logs besides the density and neutron. For instance, a similar behavior would be expected if the total porosity (PHIT) from the density (ϕ_d) were plotted against compressional velocity (V_p) or GR. For a PHIT/GR approach, assuming also that the sand laminae may contain dispersed shale but that structural shale is not present, the relevant equations are:

$$GR = (1 - V_{lam}) * (GR_{sand} + V_{dis} * GR_{shale}) + V_{lam} * GR_{shale}$$

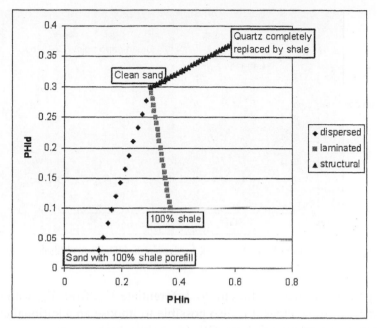

Figure 5.6.2 Thomas-Stieber Plot for Discriminating Dispersed/Laminated Shale

$$\text{PHIT} = (1 - V_{lam}) * (\phi_{csa} - (1 - \phi_{sh}) * V_{dis}) + V_{lam} * \phi_{sh}$$

where V_{dis} is the volume fraction of dispersed shale in the sand laminae, and V_{lam} is the volume fraction of laminated shale. This should lead to a plot as shown in Figure 5.6.3.

Above we have two equations in two unknowns (V_{lam}, V_{dis}), which can be solved provided that GR_{sa}, GR_{sh}, ϕ_{csa}, and ϕ_{sh} are all known or can be picked from the crossplot. The sand porosity ϕ_{sa} may then be determined using:

$$\phi_{sa} = \phi_{csa} - (1 - \phi_{sh}) * V_{dis}.$$

V_{dis} may be shown to equal

$$[1 - [(GR_{sh} - GR_0/(GR_{sh} - GR_{sa})]/(1 - V_{lam})) * (GR_{sh} - GR_{sa})/GR_{sh}.$$

In terms of known variables, ϕ_{sa} is given by:

$$\phi_{sa} = [A * (1 - \phi_{sh}) - B * (\text{PHIT} - \phi_{sh})]/[(1 - \phi_{sh}) - B * (\phi_{csa} - \phi_{sh})]$$

where $A = (GR_{sh} - GR)/(GR_{sh} - GR_{sa})$ and $B = GR_{sh}/(GR_{sh} - Gr_{sa})$.

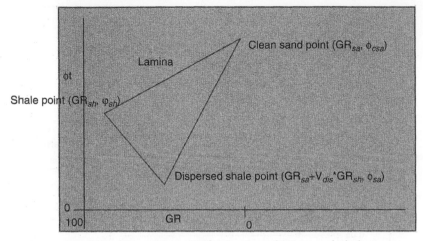

Figure 5.6.3 Total Porosity vs. GR for Laminated/Dispersed Shale System

In practice, it may be difficult to differentiate V_{lam} from V_{dis} with much accuracy. In some areas it is also possible to impose an additional empirical constraint relating GR_{sa} to GR_{sh}.

Having determined ϕ_{sa}, a conventional Archie, Waxman-Smits, or capillary curve approach may be used to determine water saturation, S_w.

Conventional formation pressure/sampling tools may be capable of identifying producible zones if one is lucky with the probe placement. Clearly it is preferred to run the tool in a packer-type mode when testing such zones. The only way to be completely sure whether a zone might be producible is through production testing. In this event I would recommend perforating the longest zone possible to give the best possible chance of encountering producible zones.

In one field I have worked in, the oil contained in missed laminated sequences was such that some blocks in the field had a larger cumulative production than the calculated STOIIP. However, when the field was reevaluated, it was found that using conventional petrophysics but removing the cutoffs that had previously been applied had the effect of more than doubling the STOIIP. In many cases it may be true that the effect of including the nonreservoir shale laminae as net sand roughly compensates for the oil volume lost from overestimating S_w in the sands (caused by the effect on R_t of the shale laminae). However, when the shale laminae are small compared with the sands, the STOIIP will tend to be underestimated. Conversely, if the zone is predominantly shale but all treated as being net, the STOIIP may be overestimated.

Exercise 5.3. Thin Beds

1. Make a Thomas-Stieber plot using ϕ_d and ϕ_n. Identify the clean sand and shale points and establish the types of clay that are present (structural, dispersed, laminated).
2. Do you consider that your evaluation in this well is affected by thin beds?
3. Also make a plot using ϕ_d and V_p. Do you learn anything additional from this plot?

5.7 THERMAL DECAY NEUTRON INTERPRETATION

Thermal decay time tools (TDTs) are used in cased holes in order to detect changes in the formation saturations occurring with time. Most commonly these changes arise from:

- Depletion of the reservoir and zones becoming swept with either water from the aquifer or injection water
- Formation of movement of the gas cap in the reservoir

The tool works by injecting neutrons, generated in a downhole minitron, into the formation. These neutrons get captured by atoms in the formation, most principally chlorine, which then yield gamma ray pulses that may be detected in the tool. Through the use of multiple detectors, the tool is able to differentiate between the signal arising from the borehole and that of the formation.

The components of the formation may be distinguished on the basis of their neutron capture cross-sections, measured in capture units (c.u.), denoted by Σ. The contractors provide charts to predict the values of Σ for different rock and fluid types. Typical values are:

Σ_m = 8 c.u. (sandstone), 12 c.u. (limestone)

Σ_{shale} = 25–50 c.u.; the value of Σ measured by the tool in a 100% shale-bearing interval

Σ_{oil} = 8 c.u.

Σ_{gas} = 2–10 c.u.

Σ_{water} (fresh) = 22.2 c.u.

Σ_{water} (200 kppm NaCl at 250°F) = 100 c.u.

The Σ measured by the tool is assumed to be a linear sum of the volume fractions of the components times their respective Σ values. Clearly the accuracy of the tool in differentiating oil and water is dependent mainly on the contrast in Σ between the oil and water. Hence the tool works well in saline environments and poorly in fresh environments. Even in a saline environment it might be found that small changes in the input parameters result in a large change in S_w. Hence the tool can give very unreliable results unless some of the water saturations are already well known in the formation.

The tool also has a limited depth of investigation, sufficient to penetrate one string of casing but not always two. It is essential to have a completion diagram of the well available when interpreting the tool, so that the relevant positions of tubing tail, casing shoes, and tops of liners are known.

Where the tool is used in time-lapse mode in a two-fluid system, clearly the variables relating to the nonmovable fluids drop out and changes in S_w can be calculated on the basis of only $(\Sigma_w - \Sigma_{hydrocarbon})$. Some of the equations that may be used for interpreting the tool will now be derived:

Two-component system without time-lapse mode:

$$\Sigma = (1 - V_{sh} - \phi) * \Sigma_m + V_{sh} * \Sigma_{shale} + \phi * S_w * \Sigma_w + \phi * (1 - S_w) * \Sigma_h \quad (5.7.1)$$

where V_{sh} denotes the total volume fraction of shale. From which:

$$S_w = [(\Sigma - \Sigma_m) - \phi * (\Sigma_h - \Sigma_m) - V_{sh} * (\Sigma_{sh} - \Sigma_m) / [\phi * (\Sigma_w - \Sigma_h)] \quad (5.7.2)$$

Two-component system in time-lapse mode:

$$\Sigma_2 - \Sigma_1 = (S_{w2} - S_{w1}) * (\Sigma_w - \Sigma_h) * \phi \quad (5.7.3)$$

From which:

$$S_{w2} = S_{w1} + (\Sigma_2 - \Sigma_1) / [(\Sigma_w - \Sigma_h) * \phi] \quad (5.7.4)$$

For a situation in which gas replaces oil with constant water saturation in time-lapse mode:

$$\Sigma_2 - \Sigma_1 = S_g * (\Sigma_g - \Sigma_o) * \phi \quad (5.7.5)$$

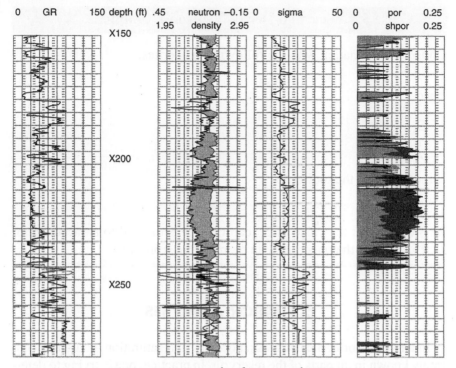

Figure 5.7.1 Example of Interpreted TDT Log

From which:

$$S_g = (\Sigma_2 - \Sigma_1)/[(\Sigma_g - \Sigma_o) * \phi] \tag{5.7.6}$$

where S_g is the new gas saturation appearing.

An example of an interpreted TDT is shown in Figure 5.7.1.

In this example, constants were chosen as follows:

Σ_w = capture cross section of water: 100 cu
Σ_{sh} = capture cross section of shale: 50 cu
Σ_{sa} = capture cross section of sand: 5 cu
Σ_g = capture cross section of gas: 7.8 cu
V_{sh} = shale volume fraction, derived from GR using GR_{sa} = 15 and GR_{sh} = 90 API
ϕ = porosity from openhole logs

Exercise 5.4. Thermal Decay Neutron Example

Consider a formation with the following properties:

$\Sigma_{sh} = 25$ c.u.

$\Sigma_{matrix} = 8$ c.u.

$\Sigma_w = 60$ c.u.

$\Sigma_o = 8$ c.u.

$\phi = 0.25$

$V_{sh} = 0.2$

Σ measured by the tool is 15.

1. What is S_w?
2. Suppose that the value of Σ_{sh} can only be estimated to an accuracy of ±5. What is the uncertainty in S_w resulting from this?

5.8 ERROR ANALYSES

In an ideal world, the net/gross, porosity, and saturation would be accurately known in all parts of the reservoir. In practice, one is trying to determine the properties based on measurements performed in a number of wells in the field, each subject to measurement error. Hence it is important to realize that there are two completely different and independent sources of error in petrophysical properties across a field. Firstly, there are errors arising from tool accuracy, sampling, and the petrophysical model, which will affect zonal averages as measured in individual wells. Secondly, there are errors arising from the fact that these properties are only "sampled" at discrete points in the field. Whether or not properties such as porosity and net/gross are mapped over the structure, or if the well data are used to make an estimate of the mean values, the result is uncertainty, which in some cases can be huge.

We will first deal with errors in the zonal average properties as measured in a particular well. I believe the most rigorous way of dealing with measurement error is through the use of Monte Carlo analysis. This method has the advantage of not requiring any difficult mathematics and is easily implemented in a spreadsheet. In this example, we will attempt to estimate the error in the average properties for a simple sand that is assumed to follow an Archie model. The basic principle is that instead of choosing point values for all the input parameters, we will allow them all to vary randomly between defined ranges, calculating the resulting values of net/gross, porosity, and saturation many times. By analyzing the resulting distributions, we can estimate an uncertainty range for each.

Starting with the net/gross, we will assume that the net sand is determined from a cutoff applied to the GR. Assume that we are using a cutoff of GR = 50 API, below which the formation is designated as sand. From inspection of the GR we may conclude that in fact the GR cutoff could have been chosen as lying anywhere between 40 and 60. Next we import the GR for the interval into a spreadsheet down one column. Over successive columns we will set the cell to 1 or 0 depending on whether the log is determined to be sand or shale at the depth increment. By summing the 1's in each column and dividing by the number of depth samples, we will determine the net/gross for a particular cutoff run.

Above the second column, the value of the cutoff chosen randomly for that run will be calculated. The formula will look something like:

$$= 50 + 10 * (0.5 - \text{rand}())$$

where rand() returns a random number between 0 and 1. Hence in the cell will be calculated a cutoff value lying randomly between 40 and 60. Let this result be denoted as GR_{co}.

In the cells below we will determine whether or not the depth increment should be designated as sand or shale. We would fill in:

$$= \text{if}(\$A2 - 2 < B\$2, 1, 0)$$

where it is assumed that the GR is in column A and GR_{co} is in cell B1.

This formula is then copied down all the cells. Below the last depth increment (line *N*) we would put in the formula:

$$= \text{AVERAGE(B2:BN)}$$

The whole of column B should then be copied to columns C, D, E, and so on, ideally at least 50 times.

Now we should have 50 separate measurements of net/gross. In another part of the spreadsheet, take the mean and standard deviation of these net/gross values. In Excel™ this can be done with the functions AVERAGE() and STDEVP(). The mean is likely to be close to that determined using a nonrandom cutoff. Two standard deviations can be used as an estimate of the uncertainty of the error in net/gross. Note that we have assumed that any inaccuracy in the GR reading itself is incorporated within the range we assumed for GR_{co}.

The next step is to calculate the uncertainty in porosity. Consider the equation:

$$\phi = (\rho_m - \text{density})/(\rho_m - \rho_f).$$

We will do exactly the same as we have done with the net/gross, i.e., import the density readings into column A of a spreadsheet. At this stage it is advisable to remove all the intervals that are designated as non-reservoir. Determine allowable ranges for ρ_m and ρ_f and at the top of each column determine the values to be used in each run. If the allowable range for ρ_m is, say, 2.65–2.67 g/cc, the equation would look like:

$$= 2.66 + 0.01 * (0.5\text{-rand}()) \text{ (denote this as RHOM, and likewise for RHOF)}.$$

Down column B insert the equation:

$$= (\text{RHOM} - \$Ax)/(\text{RHOM} - \text{RHOF}).$$

Copy column B 50 times across the spreadsheet.

Average each column at the bottom and then take the mean and standard deviation of the distribution of average porosities as you did for the net/gross. The uncertainty in the average porosity may be taken as two standard deviations. Since the porosity equation is linear, the mean porosity should be the same as that calculated through fixed fluid and matrix densities. Finally we will deal with saturation in an identical manner, although there are a few complexities. From Archie:

$$R_t = R_w * \phi^{-m} * S_w^{-n}.$$

However, we wish to finally derive a porosity-weighted saturation, so it is better to use the equation:

$$S_{wpor} = ((\rho_m - \text{density})/(\rho_m - \rho_f)) * \left\{ (R_t/R_w) * ((\rho_m - \text{density})/(\rho_m - \rho_f))^m \right\}^{(-1/n)}$$

$$(5.8.1)$$

Note that this equation is equivalent to $S_{wpor} = \phi * S_w$, substituting the porosity equation in Archie.

Above each column, it is necessary to derive randomly generated values for ρ_m, ρ_f, m, n, and R_w between allowable ranges.

At this point it should be noted that applying a range to both m and R_w is not really fair if they have been determined through a Pickett plot, since any error in one will probably be corrected by adjusting the other so that the points still go through the waterline. Hence I would apply a range to one of the two parameters only if no water sand had been available for calibration and R_w has been chosen purely from produced water samples or regional correlation. At the bottom of each column, average the S_{wpor} and then take the mean of all the runs and the standard deviation as before. The mean S_w is given by (SWPOR)average/(POR)average and the uncertainty in S_w is given by the standard deviation of SWPOR divided by (POR)average.

Finally one should have arrived at mean and standard deviations for net/gross, porosity, and saturation that fully take into account uncertainties in all the input parameters. If you are using a saturation/height relationship instead of Archie, the same process can be applied, but choose allowable ranges for your a and b values instead of m, n, and R_w. I do not believe it is necessary to take into account error in the poroperm relationships, S_{wirr}, or fluid densities, since these would be compensated for when making the $\log(J)$-$\log(S_{wr})$ plot. If, however, the saturation/height relationship is derived entirely from core, you could consider adding a term to accommodate the uncertainty in $\sigma.\cos(\theta)$.

The second stage of the process involves looking at the uncertainties in the mean values of these parameters for individual reservoir units over the entire field. At this stage it is probably useful to digress a bit and cover some elements of basic sampling theory. Imagine that one is trying to estimate the mean value of people's IQ by randomly sampling n people from a parent population of N individuals. Say the parent population has a mean IQ of M with a standard deviation of SD. The best way to estimate M is to take the mean of the IQs measured on the sample of n people, denoted by M_n. Statistical theory states that if the SD of the sample of n people is S_n, the SD of the mean of the parent population is $S_n / \sqrt{n}$.

If the accuracy of each individual IQ measurement is δ, the overall uncertainty (one standard deviation) in the value of M is given by:

$$\text{uncertainty in } M = \sqrt{((S_n)^2/n + \delta^2)}. \tag{5.8.2}$$

Hence, if we are trying to determine mean and uncertainty in the mean of, say, the porosity in a reservoir unit over the entire field, this may be estimated as follows. Take the mean and standard deviation of the various average porosities as measured in all the wells. Say these are denoted by (POR) and $(POR)_{SD}$. From the Monte Carlo analyses, one has denoted an uncertainty in the individual zonal average porosities of δ. The best estimate of the average porosity over the entire field is POR and the uncertainty in POR is given by:

$$\text{uncertainty in POR} = \sqrt{\left(\left((POR)_{SD}\right)^2/n + \delta^2\right)}. \qquad (5.8.3)$$

where n is the number of wells.

For determination of STOIIP, many parameters, including net/gross, porosity, and S_w may be input as distributions to a further statistical package that will use Monte Carlo analysis to come up with a global probability function for the STOIIP (or GIIP).

Most programs either require only a minimum and a maximum value for the parameters or require a mean, standard deviation, min, and max. In the former situation, it is recommended to take ± twice the uncertainty as calculated above as the min/max. All values within this range are treated as being equally likely. If a min, max, and *SD* are required, it is recommended to use the uncertainty calculated above as the *SD* and to take three *SD*s on either side of M as min/max, usually referring to absolute minima/maxima (zero probability of values lying outside) with something like a normal distribution about the mean. These are not the same min/max as referred to in a boxcar distribution, in which they are just ranges outside of which the values are unlikely to fall within a confidence of about 70%.

A few concluding remarks about error analysis. With regard to the Monte Carlo analyses, it is always necessary to use good judgment when considering whether the uncertainties resulting are to be considered reasonable. An experienced petrophysicist should already have a good feel for the uncertainties in the average zonal parameters he is presenting, which should roughly agree with those derived from the spreadsheets.

The sampling theory presented above assumed that n, the number of samples, is large. If this is not the case (e.g., a structure penetrated by two wells), then the results need to be treated with caution. Care should also be taken in the event that all the wells are crowded in one part of the field and there are large areas unpenetrated. This effectively means that the sampling is not random. Of course, wells are never drilled "randomly" on purpose, although looking at the actual locations of wells drilled in mature fields, they may approximate randomness rather well!

In many field static models currently being developed, all the net sand, porosity, and permeability are input from each well in the form of logs and geostatistics applied to determine the fieldwide averages, sometimes also using stochastic approaches (e.g., regarding the distribution of sand bodies). In this case the sampling part of the above becomes redundant. Such models will typically be upscaled for reservoir simulation. The saturations will then be initialized using a saturation/height function (supplied by the petrophysicist).

Sometimes it is the case that the whole model is completely wrong; for instance, based on one sample, the reservoir is assumed to be oil filled when in fact there is a gas column occupying most of the reservoir. In this case the "uncertainties" presented are obviously meaningless. It is recommended that these eventualities be considered up front and, if necessary, completely separate scenarios built up, within which the theories presented above can still be applied.

Exercise 5.5. Error Analysis

1. Copy the GR, density, and deep resistivity values from the top of the log to the OWC (oil/water contact) into a spreadsheet. For the purposes of this exercise, treat this as a single zone.
2. Use Monte Carlo analyses to determine the error in net/gross, porosity, and saturation for the averages derived in this well.

5.9 BOREHOLE CORRECTIONS

I do not wish to cover this topic in any great detail. All the contractors provide borehole, shoulder-bed, and invasion correction charts for their tools, which can be applied as appropriate.

I would like to make a few remarks about resistivity tools. Modeling of resistivity tools using analytical approaches or finite element modeling is extremely complicated. In fact no one has yet successfully modeled the combined effects of the borehole, invasion, and multiple dipping beds on the induction tool.

Chart books treat each of these effects separately. Hence there is one chart for the borehole size/salinity, another for invasion, and another for (horizontal) shoulder-bed effects. Some software is available for handling dipping beds, but these programs usually assume no borehole or invasion. In reality, all these effects combine to produce a response that is extremely complicated and will affect different tools and different depths of investigation differently. Luckily, most of the world's oil is stored in quite thick sands/carbonates. Particularly with OBM, invasion is not much of an issue, and objective hole sections are usually drilled with $8^1/_2$-in. hole, for which the borehole corrections are very small.

For quicklook evaluations, except in rare circumstances, all the necessary operational decisions, working sums, and averages can be made without the use of borehole corrections other than those automatically applied during the logging process by the contractor. For STOIIP determination, as I have stated earlier, I believe the only sensible way to go is to derive a saturation/height function. If calibrated from logs, then I have presented a means whereby the effects of invasion and thin beds can be corrected for.

Invasion as a phenomenon can actually be very useful when applied to time-lapse logging (i.e., relogging the same zone some hours or days later). For a start, the presence of invasion, as observed from a change of resistivity with time, indicates that permeability must be present. In some cases the change in properties can enable one to make conclusions regarding the nature of the formation fluid.

One area in which correction for invasion may be very important concerns modeling of acoustic impedance for seismic modeling, which will be discussed in the next chapter. In this situation, it is essential to re-create the virgin-zone sonic and density log responses from the log data, which, because of the shallow reading nature of these tools, will not be correct without proper modeling.

CHAPTER 6

INTEGRATION WITH SEISMIC

While one part of petrophysics is concerned with the interface with geol ogists and reservoir engineers in order to produce reservoir models, a further interface exists with the seismologists to ensure that the well data are used to help calibrate and understand seismic properties.

While the preparation of synthetic seismograms to tie log-formation tops with seismic horizons is a long-established technique, recent advances in far-offset seismic processing, combined with the ability to measure shear-sonic transit times, have opened up a lot of new possibilities for facies and fluid determination from seismic data. In this chapter some of these techniques will be discussed.

6.1 SYNTHETIC SEISMOGRAMS

There are two elements to synthetic seismograms. The first is the derivation of the acoustic impedance (AI) from the log data, from which reflectivity may be derived. The second is the conversion of the depth-related traces from a depth reference to a time reference so they can be compared with seismic sections. From the log data, the following may be derived:

$$V_p = 1e6/(3,281 * sonic) \tag{6.1.1}$$

where V_p is in m/s and sonic is measured in μs/ft.

$$\rho = 1e6 * density \tag{6.1.2}$$

where ρ is in kg/m^3 and density is measured in g/cc.

The AI is given by:

$$AI = \rho * V_p \qquad (6.1.3)$$

where AI is in $kg/m^2/s$.

Hence an AI trace may be derived simply from the sonic and density logs. Prior to generating the AI, it is necessary to correct for any washouts; if necessary, editing the logs by hand. It is also necessary to correct the logs for any invasion. Fluid replacement is covered in Section 6.2.

The logs should also be corrected for well deviation and datum level, such that they are true vertical and referenced to the same datum as used for the seismic survey. Having derived an AI trace in depth, it must be converted to seismic two-way time (TWT), which is the time that sound takes to reach a particular depth and to return to surface. Two sets of data are available to convert from depth to time. The first is the sonic log itself, which can be integrated to provide a total transit time. The second are data from WSTs (well shoot tests) or VSPs (vertical seismic profiles), which will give the TWT to certain depths in the well.

The normal procedure is to use the integrated sonic log to provide the conversion between checkshot or VSP points, but to calibrate the integrated sonic to honor actual checkshot points where they exist. A calibrated sonic log will provide what is known as a TZ (time vs. depth) graph, on which the TWT relating to any depth can be derived. The TZ graph is used to convert the depth-based AI log to a time-based log, denoted $AI(t)$.

The next step is usually to convert the $AI(t)$ trace to a reflectivity trace. This is simply done by differentiating the log with respect to time. The reason this works can be demonstrated as follows. Consider two adjacent samples having AI values AI_1 and AI_2. The reflectivity R is defined as:

$$R = (AI_2 - AI_1)/(AI_1 - AI_2). \qquad (6.1.4)$$

If $\delta(AI) = (AI_2 - AI_1)$ and the sampling increment is δ_t, then:

$$R \sim [\delta(AI)/\delta t] * [\delta t/(2 * AI)] \propto d(AI)/dt. \qquad (6.1.5)$$

So we can derive R from the AI by simply differentiating it with respect to time. The proportionality is not important because the trace will later be normalized before comparing with the seismic log data.

So now we have both $AI(t)$ and $R(t)$. These traces contain frequencies up to $(1/\delta t)$, and the AI also contains a direct-current (DC) component.

The next step, before they can be compared with actual seismic sections, is to convolve the traces with a seismic wavelet that is representative of the frequency content and phase of the seismic signal. Because of the nature of the seismic source and absorbing properties of the earth, the seismic survey will possess only a certain window of frequencies, somewhere between 10 and 120 Hz. The seismic wavelet is also "minimum phase" (i.e., it has a main peak occurring sometime after the event that caused it). Seismic processing can largely convert the signal from minimum to zero phase (a process called whitening) such that the wavelet is symmetric about the event with a central peak. However, it cannot replace the frequencies lost by the seismic process.

In order to make the well-derived AI(t) and R(t) comparable with the seismic log, it is necessary to convolve them with a zero-phase wavelet. Mathematically this is done by applying a zero-phase filter. One such example is a Butterworth filter (Figure 6.1.1), specified by four frequencies, such that ramps occur between the min/max frequencies and the middle frequencies, between which no attenuation is applied. After filtering, any DC components will be removed, and the traces, if plotted in a traditional "var-wiggle" format, may be compared directly with the seismic traces originating from around the wellbore. Figure 6.1.2 shows an example of logs converted to time and a synthetic AI trace.

What we have now are synthetic seismograms (commonly called synthetics). The frequency content of the seismic log may be roughly known, but it is usual to experiment with different types of filters until the char-

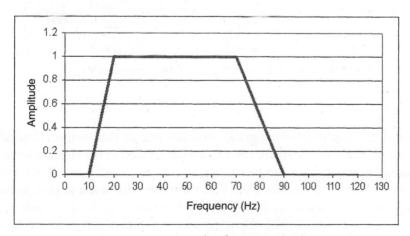

Figure 6.1.1 Example of Butterworth Filter

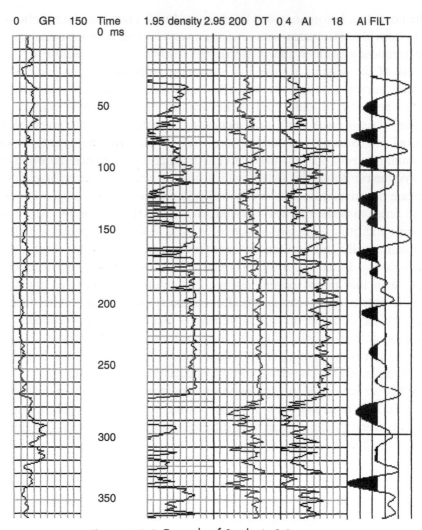

Figure 6.1.2 Example of Synthetic Seismogram

acter of the synthetic seismograms matches that of the seismic. Depending on what kind of normalization has been applied to the seismic traces, a similar normalization can be applied to the filtered AI(t) and R(t) traces to make them look similar. Note that since the AI(t) trace is now something like a sine wave, and the R(t) trace is based on the differential of AI(t) (i.e., a cosine wave), the AI(t) trace, if moved up or down by a quarter of a wavelength, will look very similar to the R(t) trace.

Big increases in AI(*t*) ("hard kicks") will result in a large peak. In the SEG (Society of Exploration Geophysicists) convention, this is displayed as a black loop (to the right). Similarly, decreases ("soft kicks") will display as white loops (to the left). A similar convention is used for R(*t*).

The depths corresponding to major changes in lithology, and hard/soft kicks, which are usually also formation tops, will have been converted from depth to time along with the AI log. These may then be overlain on the seismic to see if they correspond to seismic events. It is at this stage that the shape of the synthetic is important. Because of problems with "statics" (seismic shifts due to shallow events) or gas effects, events may have become shifted on the seismic so that they do not match the synthetic. By comparing the seismic and synthetic and matching up each loop, it may be possible to apply a static shift to the synthetic so that the two match up. Then the position on the seismic log of formation tops as seen in the well may directly tie to the seismic section.

Whether or not this works will depend on how big the AI contrasts resulting from lithology changes are, and the quality of the log data and seismic analysis. I once spent two years exclusively tying synthetics to seismic sections from logs from all around the world. In my experience the method worked very well where there were major boundaries, good-quality seismic data (usually from offshore), and reasonable log data. A near-perfect match was obtained in about 5% of cases. In about 50% it was possible to tie at least one event with confidence. It is worth remembering that given the typical frequency content of seismic (say, 70 Hz) and the formation velocities (say, 5000 m/s), one would only expect to be able to resolve events having a minimum thickness of half a wavelength, i.e., 0.5 * 5000/70 = 36 m. Below this thickness, the logs will start interfering with each other and the situation becomes much more complicated to interpret.

It should also be remembered that even if two random traces are compared with one another, they can be expected to match up about half the time. I have seen a supposedly good match proclaimed for well and seismic log data that were afterward found to have come from different countries!

When modeling a particular seismic anomaly (e.g., the pinch-out of a GOC [gas/oil contact] against the top of a structure), it is sometimes necessary to create artificial logs by successively removing a particular section of log or changing the thickness of a sand body. This can be done quite simply in the time or depth domain and the resulting traces filtered as before to create an artificial seismic section in which the effect can be modeled.

Table 6.1.1
T–Z relationship

Depth (m)	TWT (ms)
600	1000.0
620	1009.2
640	1018.4
660	1027.6
680	1036.7
700	1045.9
720	1055.1

In order to model the effect of different fluid fill on synthetics, it is necessary to use Gassmann's equations, which will be described in the next section.

Exercise 6.1. Synthetic Seismogram

1. Use the sonic and density logs to derive AI for the test1 well.
2. Use Table 6.1.1 for T-Z conversion.
3. If you have a filtering package available, apply the following zero phase Butterworth filter: 10–20–70–90 Hz
4. On the filtered AI, do you see any effect due to the OWC?

6.2 FLUID REPLACEMENT MODELING

Fluid replacement is a central part of AI modeling or creating synthetic seismograms. In essence it involves predicting how the sonic or density log will change as one porefluid replaces another. Unfortunately, the equations used to do this, developed by Gassmann, are cumbersome to apply. They also require input data that may not be readily available.

Below is presented a step-by-step menu for doing a fluid replacement, which can be applied to model the change due to any combination of water, oil, or gas with another combination. For the initial case, it is assumed that the following logs are available:

- $RHOB_{init}$ = density log, in g/cc
- DTP = compressional velocity, in μs/ft
- DTS = shear velocity, in μs/ft

Definitions:

$K_{o,g,w}$ = bulk modulus of oil, gas, water measured, in Pascals (Pa)

$K_{Finit,final}$ = bulk modulus of combined fluid measured, in Pa

K_{matrix} = bulk modulus of matrix, in Pa

U_{matrix} = shear modulus of matrix, in Pa

K_{grain} = bulk modulus of individual grains, in Pa

$S_{oi,gi,wi}$ = initial saturation of oil, gas, water (as fraction)

$S_{of,gf,wf}$ = final saturation of oil, gas, water (as fraction)

$RHOF_{init,final}$ = initial/final combined fluid density, in g/cc

$RHO_{oil,gas,water}$ = fluid density of oil, gas, water, in g/cc

Por = porosity (as fraction)

$V_{Pinit,final}$ = initial/final compressional velocity, in m/s

$V_{Sinit,final}$ = initial/final shear velocity, in m/s

V_{Pinit} = 1e6/(3.281 * DTP), in m/s

V_{Sinit} = 1e6/(3.281 * DTS), in m/s

AI_{init} = 1000 * $RHOB_{init}$ * V_{Pinit}, in $kg \cdot m^{-2} \cdot s^{-1}$

For the following, a gas/water system is assumed, but the method works equally well with oil/water.

$$K_{Finit} = 1/[S_{gi}/K_g + S_{oi}/K_o + S_{wi}/K_w] \ [K \text{ in Pa}]$$

$$RHOF_{init} = RHO_{gas} * S_{gi} + RHO_{water} * S_{wi} + RHO_{oil} * S_{oi} \ [RHO \text{ in g/cc}]$$

$$Por = (RHOM - RHOB_{init})/(RHOM - RHOF_{init})$$

$$VF_{init} = sqrt(K_{Finit}/RHOF_{init})/30.48$$

$$X_1 = RHOB_{init} * ((V_{Pinit} * 30.48) \wedge 2 - 1.3333 * (V_{Sinit} * 30.48) \wedge 2)/K_{grain}$$

$$X_2 = 1 + Por * K_{grain}/K_{Finit} - Por$$

$$K_{matrix} = K_{grain} * (X_1 * X_2 - 1)/(X_1 + X_2 - 2)$$

$$U_{matrix} = RHOB_{init} * (V_{Sinit} * 30.48) \wedge 2 \ [\text{matrix shear modulus}]$$

$$X_3 = U_{matrix}/(K_{matrix} * 1.5)$$

Matrix Poisson ratio (Mpoi) = $(1 - X_3)/(2 + X_3)$

Having determined K_{matrix} and U_{matrix}, the new V_p and V_s can be determined as follows:

$$RHOF_{final} = RHO_{gas} * S_{gf} + RHO_{water} * S_{wf} + RHO_{oil} * S_{of}$$

$$RHOB_{final} = RHOM * (1 - Por) + Por * RHOF_{final}$$

$$K_{Ffinal} = 1/[S_{gf}/K_g + S_{of}/K_o + S_{wf}/K_w]$$

$$V_{Ffinal} = sqrt(K_{Ffinal}/RHOF_{final})/30.48$$

$$Beta = K_{matrix}/K_{grain}$$

Table 6.2.1
Typical acoustic properties of fluids and minerals

Component	V_p (m/s)	K (Pa)	Density (g/cc)	Shear Modulus (Pa)
Brine	1500	2.6e9	1.05	0
Oil	1339	1.0e9	0.6	0
Gas	609	0.04e9	0.116	0
Quartz	3855	36.6e9	2.65	45.0e9
Calcite	5081	65.0e9	2.71	27.1e9
Clay	2953	20.9e9	2.58	6.85e9

$$X_4 = K_{grain} * (1 - \text{Beta})$$
$$X_5 = K_{matrix} + (1.3333 * U_{matrix})$$
$$X_6 = 1 - \text{Beta} - \text{Por} + (por * K_{matrix}/K_{Ffinal})$$
$$V_{Pfinal} = \text{sqrt}(1/\text{RHOB}_{final} * [X_5 + X_4/X_6])/30.48$$
$$V_{Sfinal} = \text{sqrt}(U_{matrix}/\text{RHOB}_{final})/30.48$$

AI_{final} may be calculated using V_{Pfinal} and RHOB_{final} as before. Typical values for constants are shown in Table 6.2.1.

Exercise 6.2. Fluid Replacement Modeling

Using a spreadsheet, model AI in the oil leg to create the response that would be expected if the well were entirely water bearing.

6.3 ACOUSTIC/ELASTIC IMPEDANCE MODELING

Gassmann's equations need to be used to correct logs to virgin conditions when making synthetic seismograms. However, they can also be used to predict the acoustic impedance of formations if the fluid changes from one type of porefill to another. Generally speaking, there are two approaches to AI modeling.

In the first approach, the AI response of the same formation, encountered with a different porefill in different wells, may be compared and also contrasted with the response of the surrounding shales. While one would expect that the water leg would have the highest AI, followed by the oil and gas legs, this is not always the case if the reservoir quality is changing between wells. Fuzzy logic techniques are usually used to fit AI distributions to the different facies types (water bearing, oil bearing, gas

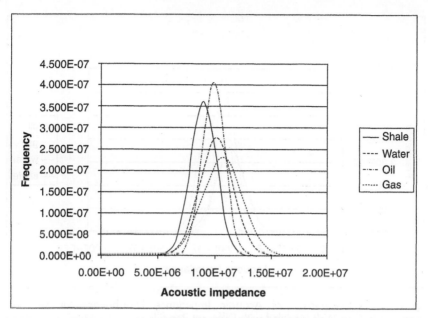

Figure 6.3.1 Comparison of Acoustic Impedance Distributions for the Same Formation Penetrated in Different Wells

bearing, and nonreservoir) and compare these to see the extent to which they overlap. They will be distinguishable on seismic only if the distributions do not overlap. Figure 6.3.1 shows an example of some distributions. In the example given, which is based on real data, it may be seen that there is extensive overlap between the distributions. Also, because the formation quality was poorer in the well that encountered gas, the mean AI for the gas sands was higher than that for the water/oil zones. This illustrates the fact that lithology effects are usually an order of magnitude greater than fluid effects.

In the second approach, one may use the formation as seen in just one well, and use the Gassmann equations to predict the change in AI as the porefill is changed. In the example given, a well that found the sand to be oil bearing was used to model the effect of changing the porefill to gas and water. Figure 6.3.2 shows the distributions. It remains the case that the sands would be overshadowed by the underlying shale distribution, although, as expected, the gas case shows a lower AI than the oil case, which is itself lower than the water case.

The fact that we got a different result depending on whether we used modeling or real well data should sound a caution to anyone using one of

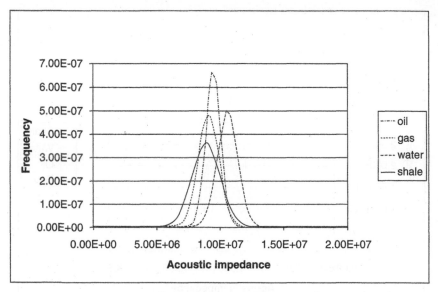

Figure 6.3.2 Comparison of Acoustic Impedance Distributions for the Same Formation in One Well Modeled with Different Porefills

these techniques to predict, say, the presence of oil-bearing sands from seismic.

While the compressional velocity does depend quite significantly on the porefill, particularly on the gas saturation, the shear velocity depends hardly at all on the fluid fill. The shear behavior of the rock may be measured with seismic to some degree by using far-offset traces. Because of the nature of sound reflecting off a surface, at high incidence angles a P wave will generate S waves when it is reflected/refracted, and these can be detected. Where only the far-offset traces are used (called AVO, for amplitude versus offset), it is possible to generate a 3-D seismic cube of elastic impedance (EI) as well as AI. The EI cube, which is mainly dependent on the shear velocity, is largely independent of fluid-fill effects.

In a similar way to generating AI traces using ρ and V_p, it is also possible to generate EI traces from the logs using ρ, V_p, V_s, and knowledge of the seismic incidence angle (θ). There are various published equations for doing this. Once such equation is:

$$EI = V_p^{\wedge}(1 + \tan^2(\theta)) * (1000 * \rho) \wedge \left(1 - 4 * (V_s/V_p)^2 * \sin^2(\theta) *\right.$$
$$\left. V_s \wedge \left(1 - 8 * (V_s/V_p)^2 * \sin^2(\theta)\right). \right. \tag{6.3.1}$$

It is possible to use fuzzy logic in the same way as with AI to also generate membership functions for various facies as a function of their EI distributions.

Hence, in the example above, it might be possible to discriminate the sands from the shales (irrespective of porefill) using the EI. Having discriminated the sands, the AI might be used to discriminate the different porefills. It should, of course, be remembered that one is still limited by the quality and resolution of the seismic. Nevertheless, in some fields, AVO techniques have provided very useful information that has led to the discovery of hydrocarbons.

Exercise 6.3. Acoustic Impedance Modeling

1. Compare the AI response in the oil leg with the AI response expected if the sand were water bearing and with that from the overlying shales.
2. Do you consider it likely that you could distinguish between sand and shale from the seismic? Between oil- and water-bearing sands? Explain.

CHAPTER 7

ROCK MECHANICS ISSUES

While rock mechanics can be a very complicated subject, there arc a few basics that all petrophysicists will need in their day-to-day work, which will be covered here. In a normal reservoir, the formation rock is subject to greatest stress from the overburden. This stress arises from the weight of rock above and can be measured by integrating the density log to surface. Since density logs are not usually run to surface, a common working assumption is that the overburden stress is approximately 1 psi/ft.

The vertical strain (i.e., compaction) caused by this stress is offset by the formation pressure, which helps "support" the rock. Because the structure is usually partially open-ended, the fluid will take only a proportion of the overburden stress. However, in overpressured reservoirs where the fluid is not free to escape, the formation pressure may become close to the overburden pressure. The net effective vertical stress seen by the formation is given by:

$$\sigma_z = P_{overburden} - P_{formation} \tag{7.1}$$

This is actually not the true effective vertical stress, which for given conditions of $P_{overburden}$ and $P_{formation}$ would result in the same strain in the sample if applied with zero pore pressure. This will now be demonstrated. Let K_m equal the bulk modulus of the matrix, when the pore pressure equals the vertical stress, defined by:

$$K_m = \text{stress/strain} = \left(P_{overburden} - P_{formation}\right)/(\delta V_m/V). \tag{7.2}$$

Let K_b equal the bulk modulus of the dry rock, as measured in a normal core measurement, defined by:

$$K_b = \text{stress/strain} = P_{overburden} - P_{formation}/(\delta V_{dry}/V).$$

$$\delta V = \delta V_m + \delta V_{dry} = P_{formation} \cdot V/K_m + (P_{overburden} - P_{formation}) \cdot V/K_b \qquad (7.3)$$

The strain is given by:

$$\delta V/V = P_{formation}/K_m + (P'_{overburden} - P_{formation})/K_b. \qquad (7.4)$$

The true effective rock stress is given by K_b* strain:

$$= K_b^*\left(P_{formation}/K_m + (P_{overburden} - P_{formation})/K_b\right)$$

$$= P_{overburden} - (1 - K_b/K_m) * P_{formation}. \qquad (7.5)$$

The factor $(1 - K_b/K_m)$ is usually denoted by α and called the **poroelastic constant**. As long as $K_m \gg K_b$, then $\alpha \sim 1$ and the net effective stress is a good approximation of the true effective stress.

This assumption may break down if soft shales are present as part of the matrix. Experiments on North Sea samples from the Fulmar formation have shown that values of α as low as 0.7 may be encountered. Where stress issues are likely to be important, such as where compaction and subsidence are likely to have an impact, it is recommended to make measurements of α on representative core samples.

For reasons that will be explained, σ_z is not the pressure that should be used for SCAL (special core analysis) measurements at in-situ conditions. Because the rock is constrained laterally, there are also lateral stresses (σ_x and σ_y), which, because of the firmness of the rock, will be less than the vertical stress. In a normal reservoir, where there is no significant difference in σ_x or σ_y, they are given by:

$$\sigma_x = \sigma_y = \sigma_z * \mu/(1-\mu) \qquad (7.6)$$

where μ is Poisson's ratio of the rock, related to V_p and V_s via:

$$\mu = [(V_p/V_s)/2 - 1]/[(V_p/V_s)^2 - 1]. \qquad (7.7)$$

The average stress (σ_{iso}) experienced by the rock is given by:

$$\sigma_{iso} = (\sigma_x + \sigma_y + \sigma_z)/3 = \sigma_z^* (1+\mu)/[3 * (1-\mu)]. \qquad (7.8)$$

In a laboratory SCAL experiment, any confining stress is applied evenly over the sample, hence "isostatic" (or "hydrostatic") conditions apply.

This means that less pressure need be applied than the σ_z calculated in the reservoir by a factor given by equation 7.4. A typical value for μ for sandstones is 0.3. This means that the stress needed in the laboratory is only $0.63*\sigma_z$. Often confusion arises about the relation between the different types of compressibility measured in the lab on rock samples, so this will be explained. First of all bear in mind that when compressing a sample by applying an external stress, the pore pressure may be either kept constant or allowed to vary. The pore compressibility at constant pore pressure, denoted by C_{pc} (= $1/K\phi$), is given by:

$$C_{pc} = -(1/V_{pore})*\partial V_{pore}/\partial P_c \qquad (7.9)$$

where V_{pore} is the pore volume and P_c is the confining stress. The bulk compressibility at constant confining pressure, denoted by C_{bp}, is given by:

$$C_{bp} = -(1/V_{bulk})*\partial V_{bulk}/\partial P_p \qquad (7.10)$$

where V_{bulk} is the bulk volume and P_p is the pore pressure. The bulk compressibility at constant pore pressure, denoted by C_{bc} (= $1/K_{dry}$), is given by:

$$C_{bc} = -(1/V_{bulk})*\partial V_{bulk}/\partial P_c. \qquad (7.11)$$

The pore compressibility at constant confining pressure, denoted by C_{pp}, is given by:

$$C_{pp} = -(1/V_{pore})*\partial V_{pore}/\partial P_p. \qquad (7.12)$$

If modeling compaction effects arising from depletion, one is typically concerned with C_{bp}, since the confining pressure (the overburden) will stay constant while the pore pressure drops. In a reservoir simulation, one is typically concerned with C_{pp}. For measurement of porosity, cementation exponent, and permeability "at overburden," one will typically supply the core contractor with the pressures to use. While the most important pressure is the one corresponding to the initial uniaxial stress conditions, measurement should also be extended beyond the pressure to cover any uncertainty in Poisson's ratio and the expected conditions at abandonment.

Note that the leak-off test, commonly conducted after first drilling out a casing shoe, may also provide useful information about the weakest lateral stress. In such a test, the formation is pressured up using the mud until close to the point at which it fractures. Hence, this pressure (after correcting for the weight of the column) will be equivalent to the weakest lateral stress. When V_p and V_s are measured in a well using a dipole sonic tool, one can directly calculate Poisson's ratio, and this is normally done as standard in the log print presented to the client.

Exercise 7.1. Net Effective Stress

You need to decide the appropriate hydrostatic lab pressure to use for some core plug measurements. Relevant data are as follows:

Depth: 12,000 ft
Overburden gradient: 1 psi/ft
Formation pressure gradient: 0.435 psi/ft
Poisson's ratio of the rock: 0.35

1. At what pressure should the measurements be performed?
2. Suppose you are now told that the poroelastic constant for the samples is 0.85. What pressure should you use?

VALUE OF INFORMATION

It is important for petrophysicists to have a feel for the economic impact of the work they are performing and whether or not the cost of running a certain log is really justified in view of the economic benefit it will ultimately bring. In this chapter, some considerations and tools for assessing this will be explained.

In a normal oilfield economic model, money is expended on exploration until a discovery is made. Following discovery, there is a development phase involving significant capital expense (called CapEx) on wells and facilities. At a certain point, money will start coming in from production and start to pay back CapEx. There will also be ongoing operating expenses (OpEx) and tax on revenues. At the payback time, the revenues will have covered the sunk CapEx and OpEx and the project starts to move into the black. At any point in time, the field will have a future value (ignoring all the sunk costs), which will be denoted as net present value (NPV). The NPV will be calculated from the production forecasts, together with assumptions about hydrocarbon prices, taxes, and future OpEx and abandonment costs. The time element in these costs and revenues is taken into account with present value accounting, which relates all cash flows to a fixed reference point.

It is important to realize how information is related to NPV. Obviously the more information you have about a field, the more wisely the CapEx may be expended (e.g., in right-sizing the facilities and drilling the most cost effective wells) and the greater the revenue. However, there will be an effect of diminishing returns. This is illustrated in Figure 8.1. Due to the cost of information, the NPV will rise with information up to a point, then start to fall. Even if money is being spent on information at a steady rate, the incremental value will become less with time during the life of

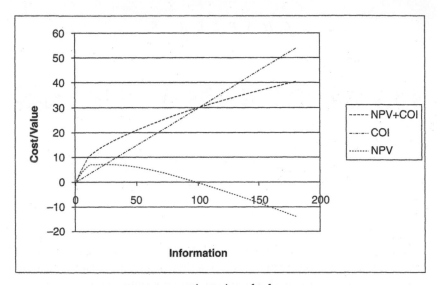

Figure 8.1 The Value of Information

the field, since there are no decisions left to be made that can lead to greater revenue on the basis of the information. An example would be acquiring a core in a field one month before abandonment—the data cannot be used to change anything, so the money is just wasted.

At this point it is important to get a feel for the relative amounts of money one is talking about, at least with respect to the impact of logging. Say that very early in the life of an assumed 50-MMbbl field, prior to designing any facilities, it is decided to include nuclear magnetic resonance (NMR) logs in all the early development wells. This comes at a cost of half a million dollars, but it is assumed that the tool always gives correct results. The logs are justified because of known concerns about the conventional evaluations, since it is considered that there is a 30% chance that the stock tank oil initially in place (STOIIP) is being seriously underestimated and could be as high as 75 MMbbl.

Consider what will happen if the logs are not run. The $500,000 will not be spent. There is a 70% chance that the facilities will be designed correctly and that the field will realize an NPV of, say, $500 million. There is, however, a 30% chance that the STOIIP is in fact 75 MMbbl. If this is the case, then the facilities designed for 50 MMbbl will be suboptimal and will result in deferred production and slightly less ultimate recovery factor. The economic impact of this would be such that the NPV would be only $650 million.

If the facilities were to be right-sized for 75 MMbbl, the NPV would be $700 million. The estimated additional monetary value (ΔEMV) of running the NMR logs can be calculated by:

$$\Delta\text{EMV} = (0.3 * 700 + 0.7 * 500 - 0.5)$$
$$- (0.3 * 650 + 0.7 * 500) = \$14.5\text{M}. \qquad (8.1)$$

Obviously in this case the decision to run the NMR logs is expected to make a profit. However, note that the decision to run the NMR logs has not gained you 25 MMbbl oil, which might have a value of $500 million. It has only allowed you to make decisions that have made you develop the field more efficiently. Most of the extra 25 MMbbl would have been produced anyway.

Up until now you have also assumed that the tool always leads you to the right result. Consider instead the situation in which you run the tools but have a confidence of only R (expressed as a fraction) that they will give you the right answer. Here you have opened up the possibility of building facilities for a 75-MMbbl field that is only in fact a 50-MMbbl one. In this event, say you make only an NPV of $400 million. These data may be put in the form of a decision tree (Figure 8.2).

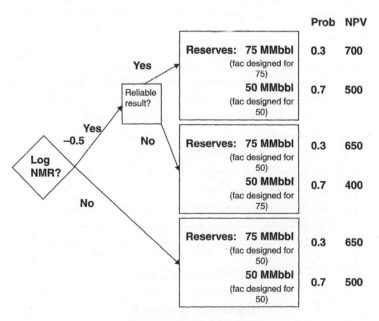

Figure 8.2 Decision Tree

Following the same logic through as before, the EMV is:

$$\Delta EMV = (R*(0.3*700+0.7*500)+(1-R)*(0.3*650+0.7*400)-0.5)$$
$$- (0.3*650+0.7*500).$$

(8.2)

For the case in which $R = 0.5$ (50% probability of being correct):

$$\Delta EMV = 517 - 545 = -28$$

Hence, your decision to undertake NMR logging has cost the company $28 million!

Whenever I have seen these kinds of decision-tree calculations performed, the possibility that the data acquired may lead one to make the wrong decision is never considered. Equation 8.2 allows you to calculate the value of R at which the data acquisition becomes worthwhile. If we plot the ΔEMV vs. R, we get the result shown in Figure 8.3. From the plot we can see that below a reliability of 67%, the NMR tool is not worth running. The plot can also tell us what the economic benefit would be of taking steps (e.g., further tool calibration, special studies, etc.) to improve the reliability of the tool. Such plots are a persuasive means of convincing management of the benefits of data acquisition

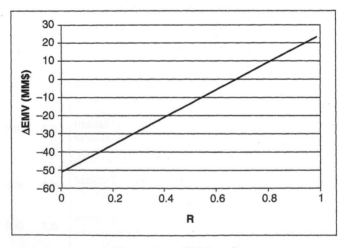

Figure 8.3 ΔEMV vs. R

or research campaigns. The mathematical concepts in the probability theory concerned with value of information (VOI) are discussed in Appendix 4.

Generally speaking, early in the life of a field the EMV of correct information is very high. However, the negative EMV of information that is misleading may be generally even higher. This is because the loss in NPV arising from making your facilities too small is much less than the cost of making them too large—the former will most likely only cause deferment, while the latter will see unused capacity that is just lost capital. On top of that, you had to pay for the information in the first place.

Late in the life of a field, the situation is completely different. You may be largely stuck with the facilities you have, so finding out that your STOIIP is 75 MMbbl instead of 50 MMbbl will not lead to any major change in your development—you will just make more money than you originally expected to. Likewise, if you find out that your STOIIP is less, there is nothing you can do about it. In essence, the value of the information becomes much smaller, while the acquisition cost remains the same. Hence, whereas in the initial example the cost of the logs was effectively negligible but the reliability was the crucial issue, at the end of field life the cost may be a major factor but the impact of unreliable data is relatively less.

The golden rules for deciding whether a course of data acquisition is justified may be summarized thus:

- When considering the potential economic benefits from the data, you have to consider the whole field economics, and crucially what different decisions will be made as a result of the outcome, and what their impact will be.
- Remember that "finding" additional hydrocarbons does not necessarily have an economic value equal to the spot price of those hydrocarbons.
- You must factor in the probability that the information will lead you to make the wrong decision as well as the right one.
- Remember that there is always the option to acquire the data but to choose not to act on it if you consider it to be unreliable or if it just confirms what you already assumed. Certain data (e.g., virgin formation pressures) can be acquired only at an early stage in the field development, so it is sometimes better to acquire them even if their impact cannot immediately be assessed.

Exercise 8.1. Decision Tree Analysis

Consider the following scenario:

You are trying to decide whether to run a thermal decay time (TDT) tool in an old producing land well. Log costs are $20,000 (including cost of mobile wireline unit, lubricator, etc.). The well is currently producing 200 barrels of oil per day (bopd) from zone A in the well. With no intervention it is anticipated that the well will produce a further 60,000 barrels. However, zone B, which was never perforated in this well, may still be producible. There is a 50% chance that it is already flushed, but if it is not, then it might be expected to produce 600 bopd a year, or a total of 180,000 barrels.

After running the TDT, you have the option to decide whether to leave things as they are or to spend $1 million doing a workover, which will entail abandoning zone A. The formation water is not very saline, and there is no base TDT log to compare with. Therefore, reliability of the results is estimated at only 70%. If you were to recomplete the well, believing zone B to be not flushed when in fact it was, the well would have to be abandoned.

For the purposes of the VOI exercise, assume an oil value of $20 per barrel (after tax).

1. Is the TDT justified? If not, at what level of reliability would the tool run be justified?
2. Repeat the calculations for the situation in which there is a 30% chance that the zone is already flushed, and for a 70% chance.
3. Consider the case that there is a 50% chance that the zone is flushed, but while the chance of the tool correctly indicating a nonflushed zone to be nonflushed is 70%, the chance of the tool correctly indicating a flushed zone to be flushed is only 60%.

EQUITY DETERMINATIONS

Equity determination becomes necessary when part of an accumulation extends across a boundary and becomes subject to different conditions. Such a boundary may be a result of (a) different ownership of the acreage, such as occurs in the North Sea, where different groups of companies control different blocks, or (b) international boundaries. In either of these cases, it becomes necessary to determine the relevant amounts of hydrocarbons lying on either side of the boundary. Typically, following an equity determination there will be a **unitization agreement**, whereby a single commercial unit is formed with the aim of optimizing the total recovery of the field.

Equity denotes the share of this controlling unit held by the various parties. A higher equity will involve a greater share of the profits, but also a greater share of the costs and liabilities. Since the parties involved will want to make the most profit from the field, there will usually be an attempt by each party to maximize its own equity. The process whereby an agreement is reached on how the equity is divided is called an **equity determination**.

Especially where international boundaries are concerned, equity determinations may take many months or years to conclude and may involve significant deferment of the hydrocarbon production. Recognition of this fact, together with a tendency for field sizes to become smaller, has led to a more pragmatic approach in recent years. However, the costs of the technical work are still sizable. When a field is first discovered and there are insufficient data available to make a proper equity determination, "deemed equity" will typically be agreed upon by the parties. This is a rough working agreement to enable appraisal/development of the field to progress, with costs reallocated and recovered as appropriate following a full equity determination at a later stage.

9.1 BASIS FOR EQUITY DETERMINATION

Clearly there are different ways to determine equity, based on the parameters likely to define the ultimate value in the field and what can easily be measured. These are:

1. **Gross bulk volume** (GBV). It might be argued (by the party with the greatest GBV) that the fairest way to divide equity is on the basis of the total bulk volume of hydrocarbon-bearing rock between certain horizons. This would have the advantage of being relatively quick and simple to determine, since the parties would have to agree on only the seismic interpretation and mapping procedures. However, the other parties would rightly argue that it is not the rock that has the value, but only the hydrocarbons that are produced from it. Therefore, as long as reasonably reliable methods are available to determine the hydrocarbon production from either side of the boundary, to make a determination purely on the basis of GBV would be unfair.

2. **Net pore volume** (NPV). This determination might be justified on the grounds that it is also relatively simple to determine. However, if the reservoir quality on one side of the boundary is relatively poor compared with the other, the contribution to production arising from the poorer side will be relatively less. Also, if one side has a relatively low relief above the free water level (FWL), then it will contain relatively less hydrocarbon. Such a determination also makes no distinction between the relative value of oil and gas.

3. **Hydrocarbon pore volume** (HCPV). By introducing the saturation, some of the drawbacks of an NPV basis are removed, but this would still not take into account the fact that not all the hydrocarbon may be recoverable (e.g., if it is located in very thin or low-permeability rock), as well as the relative value of gas and oil.

4. **Barrels of oil equivalent** (BOE). For this basis, a factor is used to convert in-situ gas volumes to an equivalent oil volume on the basis of value. This factor will reflect the local value of gas compared with oil, bearing in mind the costs of transporting it. Hence it takes into account the fact that one side may have relatively more gas than the other side. Note that there are two sorts of gas to be included: the associated gas that is produced from the oil and any free gas existing in the reservoir. Typically the same gas factor will be used for both.

5. **Reserves**. It might be argued that this is the only fair way to properly allocate value. However, in practice it may be impossible to agree on

how much actual production arises from the different sides of the boundary. The relative contributions will necessarily be a function of the development strategy and positions of the wells.

The technical teams working in each company will typically make their own estimate of equity on all of the above bases. Having determined which one is most favorable, they will try to propose that basis to the other parties. Needless to say, one party will likely propose a particular method only because it suits it, not out of any genuine desire to "keep things simple." In all equity determination I have been involved in, the final basis has been on BOE. This is usually the most logical basis for determination.

9.2 PROCEDURES/TIMING FOR EQUITY DETERMINATION

As stated above, at the early stages of field development, the senior managers of each company will often get together and agree on a deemed equity to allow development to commence. An initial deemed equity that is low will mean that the cash contribution required for the development will be less. Therefore, it might be considered advantageous to minimize one's deemed equity. It may often be the case that the final equity does not differ much from the deemed equity. Some companies are thus wary of pegging their deemed equity at a low value because they believe it will tend to result in a lower ultimate equity.

During the deeming phase, parties may not necessarily share the same data or make well data they have acquired on their side of the boundary available to other parties if they feel it will be detrimental to their case. The managers also have to agree on who will operate the field and the infrastructure. In some cases, sunk well costs might be shared between the parties if they can be used for development purposes. All these considerations will be incorporated within a **joint operating agreement** subject to an interim unitization agreement.

Generally speaking, the deemed equity will remain in place during the development drilling until some time around first oil (or gas). At that point there will be a provision for one or all parties to request a full equity determination. Note that it is not essential that this ever actually occur. Each party will consider how much it is likely to gain (or lose) from such a determination, also bearing in mind the costs of it. It might be that all parties, rightly or wrongly, do not consider it to be in their interests to call a full equity determination, in which case the deemed split may remain in force. There will typically be provisions made for further equity determinations during the life of the field. However, as time progresses, the incentive for determination may be less in view of the decreasing remaining value of the field and likely increasing determination costs (resulting from additional data acquired).

Many methods have been used for the actual determination. Among the most common are:

1. **Technical determination.** In this procedure, a series of technical meetings are called between the parties, and in theory a common technical position is agreed upon from which the equity may be determined. In practice these have typically been found not to work, since each party will tend to cling to a technical case that optimizes its own position. Such negotiations have been known to go on for years, with eventual resort to the courts, resulting in a loss of value all around. Therefore, this procedure has more or less been abandoned (and rightly so).

2. **Fixed equity.** Early in the field life, a company that is likely to control a very high proportion of the equity may offer the minority parties a fixed percentage, which is not subject to future equity determination. This has the advantage for the majority shareholder of not having to devote a lot of resources to negotiating over a small percentage or even share data with the minority party. It may have advantages for the minority shareholder that is a small company and does not want the expense of a full equity determination.

3. **Management negotiation.** This is just an extension to the deemed equity procedure. The managers might recognize that some adjustment is needed to the deemed equity after some years, but they do not wish to go to the full expense of a full technical procedure. They would normally give themselves the option to call a full equity determination at some point in the future.

4. **Sealed bids.** Full technical determinations can also be avoided if each party makes a sealed bid, stating its equity case, to an independent third party (such as a judge). If the sum of the equity bids is less than, say, 110%, each share can be prorated downward accordingly. If the total is greater than 110%, then a full technical procedure can be instigated.

5. **Expert determination.** In a full technical procedure, all of the available data are handed over to an independent expert who will make the determination. In theory this sounds like the most logical way to do an equity determination. However, a number of pitfalls can arise:
   • The parties have to be able to agree on a suitable expert.

- Essentially all the technical work has to be reperformed, involving considerable time and cost. There may also be difficulties in transferring all the data to the expert in a format that can be readily used.
- If the expert's results significantly erode the equity of one party from that previously assumed, the results may be challenged and not accepted by the party disadvantaged.

6. **Expert guidance.** Rather than have the expert perform all the technical work from scratch, another option is for all the parties to submit a technical report to the expert, stating their respective cases for their proposed equity positions. By auditing the reports and analyzing the strengths and weaknesses of each, the expert may derive a new equity split. This is certainly a quicker and cheaper alternative to expert determination. However, the problem remains that a party losing equity may challenge the expert's results.

7. **Expert pendulum arbitration.** This procedure is similar to expert guidance, except that the expert is asked only to choose between the various cases submitted, not interpolate between them. The idea is that it puts pressure on the parties not to make a claim for unreasonably high equity, since it will probably result in their case not being chosen. This procedure is quicker and simpler than expert guidance but suffers from the same weaknesses. It also introduces a "lottery" element that some companies might not find acceptable.

In my view a management negotiation is nearly always the preferred route for equity determination, since it saves a lot of time and money, and even a full technical determination is probably not nearly as accurate as the technical experts would have us believe. However, in most of the equity determinations I have been involved in, the goodwill between companies has been lost at an early stage as a result of middle managers trying to gain points and taking a nonpragmatic view. Therefore, there is usually no alternative to a full equity determination. I am aware of one company devoting 16 man-years of work in-house to an equity determination, with additional expert/legal costs amounting to millions of dollars. I also know of a field (straddling an international boundary) that was shut in for 7 years because a unitization agreement could not be reached.

9.3 THE ROLE OF THE PETROPHYSICIST

Assuming that a full technical determination is made that is related to HCPV or BOE, the petrophysicist will play a crucial role in the equity case that can be made on behalf of his company. Clearly the petrophysi-cist has a duty to calculate parameters using accepted and correct method-ologies. However, it is perfectly reasonable that the methodologies should be chosen and applied in such a way that they happen to provide the most favorable equity position for his company.

It is not always obvious how the equity will be affected by a differ-ent choice of methods and parameters. Therefore, the first step is to set up a model that will enable the effects of changing parameters and models on the equity position to be determined quickly. This is most easily done as follows. Choose a well on either side of the boundary that may be considered reasonably representative of the existing wells. In some cases, it might be necessary to choose more than one well, particularly if both gas and oil are present. Copy all the raw log data into a spreadsheet. Set up the following parameters as global variables:

- V_{sh} cutoff
- Porosity cutoff
- Grain density
- Fluid density, gas leg
- Fluid density, oil leg
- Fluid density, water leg
- R_w
- m
- n
- FWL
- Gas/oil contact (GOC)
- J function, S_{wirr}
- J function, a value
- J function, b value
- J function, $\sigma\cos(\theta)$ (oil/water)
- J function, $\sigma\cos(\theta)$ (gas/water)
- Oil density
- Gas density
- Water density
- Gas factor (for converting in-situ gas volume to oil volume)

Set up the evaluation of the wells in terms of the above variables so that you are able to provide proxies for the equity variables for each well as follows:

- GBV: from the height of the total hydrocarbon-bearing column
- NPV: from the sum of the net footage times the porosity
- $HCPV_1$: from the sum of the net footage times the porosity times $(1 - S_w)$, S_w derived by Archie
- BOE_1: as per $HCPV_1$ but multiplying any gas footage by the gas factor before summing
- $HCPV_2$: as per $HCPV_1$ except using the J function instead of Archie
- BOE_2: as per $HCPV_2$ but multiplying any gas footage by the gas factor before summing

Let the well on your side of the boundary be denoted by A, and the one on the other side by B. Determine the following parameters:

$$EQ(GBV) = GBV(A)/(GBV(A) + GBV(B)),$$

where GBV(A) is the GBV proxy determined from well A, etc.

$$EQ(NPV) = NPV(A)/(NPV(A) + NPV(B))$$

$$EQ(HCPV_1) = HCPV_1(A)/(HCPV_1(A) + HCPV_1(B))$$

$$EQ(BOE_1) = BOE_1(A)/(BOE_1(A) + BOE_1(B))$$

$$EQ(HCPV_2) = HCPV_2(A)/(HCPV_2(A) + HCPV_2(B))$$

$$EQ(BOE_2) = BOE_2(A)/(BOE_2(A) + BOE_2(B)).$$

Note that if it has been decided to use more than one well on either side of the boundary, the equations can easily be extended, using relative weighting factors as appropriate. For example, if two wells are chosen for your side (A and B) but only one is available from the other side (C), and you feel that well A is likely to be twice as important as well B, you could calculate:

$$EQ(BOE_1) = (BOE_1(A) + 0.5 * BOE_1(B))/$$
$$(BOE_1(A) + 0.5 * BOE_1(B) + BOE_1(C)).$$

We are not interested in the absolute value of $EQ(BOE_1)$, only its relative size compared with other proxies for equity (e.g., $EQ(BOE_2)$) and how it varies subject to input petrophysical parameters. Which of these is the maximum will give an indication of which basis for determination is likely to be the most favorable to your company. However, we will assume for further discussions that the BOE method is decided upon as being the most reasonable basis for equity determination.

At this point it is probably best to decide whether it is going to be advantageous to push for a saturation/height function approach to saturation determination or to opt for a simple Archie approach. Arguments that can be used in favor of either approach are:

Pro Archie:
- It is relatively easy to agree on the Archie input parameters, particularly if SCAL (special core analysis) data are available and the water leg has been logged.
- Most petrophysicists will tend to average capillary (cap)-curve data a different way, and it is hard to agree on a common saturation/height function.

Pro *J* Function:
- Saturation/height functions are the only acceptable way to do volumetrics in a field.
- This is the type of function that will be used for the dynamic model, so it should also be used for the volumetrics.

The next step is to determine the relative weight that each of the various input parameters has on the equity. This is done by varying the input parameters sequentially within justifiable ranges and observing the effects on $EQ(BOE_1)$ and $EQ(BOE_2)$. The result might be as shown in Figure 9.3.1.

These results tell you where you can hope to make the most impact on equity. If negotiating which parameters to use with partners, you can decide up front where you should make a strong case for a particular parameter being used, possibly in exchange for being more lax on another parameter.

Net/Gross

If the wells on your side of the boundary tend to have formations of poorer quality than those on the opposing side, it is in your interest to push for little or no cutoffs to be applied. As stated in Chapter 2, there are many good arguments for not applying cutoffs, but even those that are applied can be calculated in a way that is favorable to your side through

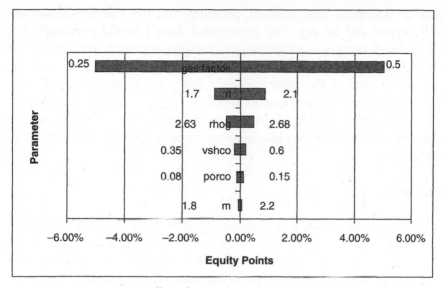

Figure 9.3.1 Effect of Petrophysical Parameters on Equity

the way that V_{sh} is calculated and the V_{sh} cutoff that is applied. If the wells on your side are relatively good, you could make an argument that for equity, as distinct from normal calculations of HIIP (hydrocarbons initially in place), it is only reasonable to include sands that are likely to contribute to production via the base-case development scenario, rather than to include those that could conceivably produce, given some exotic recovery mechanism yet to be devised.

If using a V_{sh} cutoff results in lower equity than using a porosity cutoff (based on the density) or density/neutron crossover, then you can make a case for using an alternative method. If net/gross is a property that will be mapped over the reservoir, you need to further consider how this mapping will be done. Within commercial contouring packages, there are different algorithms that can be used for this that will give different equity results. You might also consider it in your interest to propose just using a constant net/gross over the entire field or using a different constant on either side of the boundary.

Porosity

Assuming that the density log method is being used to determine porosity, you can investigate which choice of fluid density and grain density optimizes your position, provided that the values you use are still supportable by any core or regional data. I would personally not recommend using a density/neutron crossplot approach even if it gave a better equity case. However, this method is widely used in some oil companies, and if the other company proposes it (and it favors your equity position), then you might not choose to protest too strongly against the method being adopted. As with net/gross, how the mapping of porosity is performed may also have an impact on equity and needs to be considered.

Saturation

As discussed above, it is recommended to agree early on whether a resistivity-derived or saturation/height approach should be used. If an Archie model is adopted, then you can try to optimize your equity through the choice of R_w, m, and n. As before, any values proposed should be supportable by either core or log data.

If appropriate, a shaly sand model may also be applied. However, the more complicated the methodology becomes, the harder it will be to reach an agreement with other companies as to the parameters that should be used. If it is agreed to use a saturation/height function approach, then the parameters defining the curve that optimize equity can be proposed in a similar way as with Archie. Since the proper methodology for averaging cap curves is not always well understood by petrophysicists in oil companies, by proposing a function up front, supported by the available cap-curve data, one has a good chance of getting it accepted.

Fluid Contacts

Since the reservoir shape will usually not be symmetric with respect to the boundary, the position of the contacts will often have a large effect on equity. Note that if a saturation/height function is being used, the FWL should be used as the cutoff for volumetric determination. If a conventional Archie approach is being used, then the hydrocarbon/water contact (HWC) is more appropriate. For reservoirs with a gas cap, moving the GOC up or down will either favor or disfavor one company.

Where contacts are clearly observable on the logs, there is not much room for debate. However, where this is not the case, then there may be a wide range of possible contacts, depending on assumptions made with regard to the formation pressures, spill point, bubble point, etc. As before, you need to propose the model, supported by at least part of the data, that optimizes your position. If a two-well model is being used to estimate the effect on equity, care needs to be taken that this is indeed representative. For complex reservoir geometries where additional accumulations open up as the HWC is deepened, it may be necessary to go to a full mapping package to properly assess the impact on equity of moving the contact one way or another.

Exercise 9.1. Optimizing Equity

Consider the following scenario:

Well test1 is on your side of a block boundary with well test2, which is part of the same accumulation, on the other side of the boundary and controlled by another company. You have agreed to share data with the other company and to pursue a common petrophysical model for evaluating the accumulation. The test2 well has no core data.

You have agreed to give a technical presentation to the other company with your recommended model for the evaluating the field. This model, if accepted by the other company, may be used as the basis for an equity determination.

Perform an evaluation of the test2 well. Fully detail a technical defensible model for the evaluation of both wells which optimizes your equity position.

The log data from the test2 well is in Appendix 3. Assume that the data are true vertical relative to mean sea level.

CHAPTER 10

PRODUCTION GEOLOGY ISSUES

The purpose of this chapter is to provide readers with a convenient reference for production geology as it relates to the petrophysicist's daily tasks, though not intended to be a comprehensive guide to the wider discipline. The interface between the petrophysicist and production geologist is crucial in ensuring that:

- Wells are proposed and drilled in optimum locations
- The correct operational decisions are made while the well is drilled
- The field model makes optimal use of the available well data
- Production from the well can be properly understood in a structural context

For this interface to work well, it is essential that both discplines have a working understanding of each other and a common terminology.

The primary duties of the production geologist are as follows:

1. Correlate all the available well data within the production area, providing a logical and consistent formation designation scheme.
2. Prepare, in conjunction with the seismologist, geological subsurface interpretations comprising subsurface maps of key horizons and cross sections.
3. Update subsurface interpretations as new well/seismic data become available.
4. Advise on selection of new well locations.
5. Determine the gross bulk volume (GBV) of the reservoir, which may be used, in conjunction with data provided by the petrophysicist, to determine net pore volume (NPV), hydrocarbons initially in place

(HIIP), etc. Such models may be either deterministic (i.e., using fixed distributions of reservoir properties) or probabilistic (i.e., using probability distributions for reservoir properties).

6. Create an upscaled subsurface model in digital format that may be exported to the reservoir simulator and used for dynamic simulation and reserve estimation.
7. Provide the geological background for any proposed well stimulation or secondary recovery projects.

Poor communication between the petrophysicist and the geologist can sometimes lead to some expensive mistakes, particularly with respect to proposing future wells and assessment of reserves. Here are some of the major pitfalls I have come across:

- Where the reservoir zonation is quite coarse (i.e., one zone covers a large depth interval), there may be considerable variation of the reservoir properties over the zone. One petrophysical average may be quite inappropriate. For instance, say a zone consisted of 3 m of good-quality sand of 1 darcy permeability overlying a 100-m interval having 10-md sand. The zone, seen as an average, might appear unproducible when actually the good sand, taken in isolation, might be commercial. It could also happen that the whole zone is interpreted as commercial when in practice only a small part will ever contribute to production.
- Where the reservoir zonation is too fine, there is a danger of incorrect correlation between wells. This often leads to an incorrect assessment of fluid contacts or to some serious errors in estimation of reserves.
- Great care needs to be taken in the allocation of permeabilities in the production geologist's model. Bear in mind that this may be done in a number of ways. The petrophysicist may apply a poroperm equation to his porosity log and supply the production geologist with a curve to use within his software. In this case the petrophysicist needs to ensure that the permeabilities in nonreservoir units are set to an agreed-upon value. In particular, very high permeabilities arising from spurious porosity values need to be edited out.
- The petrophysicist may supply the production geologist with a poroperm equation to apply himself. In this situation there is an even greater danger of incorrect values entering the model. If the geologist is resampling the data to a coarser depth interval before applying the equation, the resulting permeabilities are very likely to be incorrect due to the nonlinearity of the poroperm equation.

- The petrophysicist may supply the geologist with a constant average permeability to use for the horizon in question, or supply averages for each well. There is a danger here that either the average does not take into account known geological variation over the structure or that contouring of well averages leads to an incorrect interpretation of the areal variation.
- The above arguments with respect to permeabilities may also apply to water saturations. The petrophysicist may supply the production geologist with saturations in the form of either logs, averages, or a saturation/height function. It is essential that a clear audit trail for the saturations in the model be supplied.
- Probably the greatest source of error in the petrophysics/production geology interface lies in the realm of net-to-gross values, and this has led to some huge mistakes in the past. Probably the safest approach, where the petrophysicist is supplying the geologist with evaluated logs for inclusion within a static model, is for porosity to be set to zero in all nonreservoir units, and net/gross to be set to unity throughout. However, even this can cause problems where upscaling is occurring. It is really essential for the petrophysicist to sit with the geologist and see just how net/gross is being handled within the software used to generate the static model and how this is passed on to the dynamic model.
- Picking of coring points or well TD (total depth) is often done by the production geologist on the basis of his correlation. Failure to incorporate all of the petrophysical information available may often result in bad decision making.
- Sometimes the petrophysical interpretation itself depends on the geological interpretation. For instance, if gas/oil differentiation is not possible from the logs alone, use might be made of the known production history from neighboring wells, which will depend on the correlation. If this correlation is wrong, the fluid allocation will be wrong. It may also happen that the petrophysicist makes an interpretation of the fluids that leads to an unresolved inconsistency between wells. In this case the production geologist may be forced to introduce a fault in the structure, which may or may not exist in reality.
- The logging program may incorporate items that have a positive value of information (VOI) only in the event of the well being a success. Therefore, the petrophysicist, using good communication with the production geologist, may be able to save money on the well through provision of early information on the well's results. Poor communication

will often lead to logs being run that have no value. Conversely, where the results are unexpected, additional logging having a high VOI may also be proposed.

- The petrophysicist and production geologist will often be using well deviation data from different sources. It is essential that these be checked for consistency before any work on true vertical log data is shared.
- Both the petrophysicist and the production geologist may have access to reports and logs that are outside the domain of information shared digitally between departments. Where there is poor communication and lack of a proper library structure, it may often occur that neither has access to the most complete information that can be used to improve his models. On numerous occasions I have seen this with respect to core data.

10.1 UNDERSTANDING GEOLOGICAL MAPS

10.1.1 Basic Concepts

Consider a three-dimensional surface, such as the top of a particular horizon in the subsurface. If you were standing on such a surface, there would be a direction in which the surface slopes most rapidly. Relative to north, this direction would have an azimuth, referred to as the **azimuth of the dip direction**. The angle between this direction and the horizontal is referred to as the **dip magnitude**. If we were to take a horizontal line perpendicular to this direction (called the **strike line**) and measure the angle going clockwise from north to this line, we would have the **strike direction**. These items are illustrated in Figure 10.1.1.

The strike lines of the surface, when combined for a specific horizontal elevation, form **contours**. Maps of a surface are created by showing contour lines for fixed vertical spacing. For a smooth surface, these lines will be continuous. However, where the surface is not smooth (for instance, where faulting occurs), the lines will be discontinuous. The average dip magnitude may be measured from a contour map by taking the distance between contour lines and using the formula:

$$\tan(\alpha) = (\text{vertical contour spacing})/(\text{horizontal contour spacing}) \quad (10.1.1)$$

where α is the dip magnitude. Example: If the contours on a $1:50,000$ map are every $200\,\text{m}$ and are spaced by $6\,\text{cm}$, what is the dip magnitude?

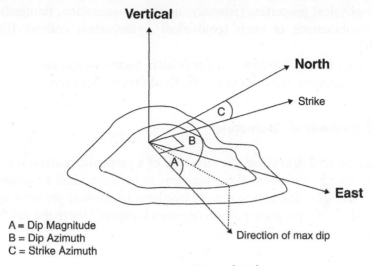

A = Dip Magnitude
B = Dip Azimuth
C = Strike Azimuth

Figure 10.1.1 Dip and Strike

$\tan(\alpha) = (200)/(50000 * 0.06) = 0.067$

$\alpha = 3.8$ degrees.

Exercise 10.1. Dip Magnitude

If the contours on a 1:25,000 map are every 100m and are spaced by 4 cm, what is the dip magnitude?

10.1.2 Types of Maps

The map described in the first paragraph of the preceding section concerns the depth of the top of a particular horizon and is therefore familiar to anyone used to reading geographic maps. However, mapping does not have to be limited to just the parameter of depth. The technique can be used to represent any parameter that varies areally over a structure. Other parameters that may be mapped and contoured include:

- Thickness of a particular horizon: This may be either the isochore thickness (i.e., in the vertical direction) or isopach thickness (in the direction normal to the bedding plane).

- Petrophysical properties (porosity, net/gross, saturation, permeability) or combinations of such (equivalent hydrocarbon column [EHC], NPV)
- Paleography, environments of deposition, facies variations
- Fluid properties (depth of contacts, fluid density, salinity)

10.1.3 Methods of Contouring

Based on well data alone, the structure of a particular horizon is known only in discrete locations. What happens in between must be estimated. Various mathematical algorithms are available to estimate the most appropriate values of a parameter to use between locations where the values are known with certainty. These are:

1. **Triangulation.** Straight lines connecting data well locations are created, and along these lines the data are interpolated linearly. The subsequent contours may be smoothed using a "spline fit," which attempts to reduce the second derivative of the curve.
2. **Inverse distance.** In this technique, the inverse of the distance from each known data point is used to establish a weighting to use for taking an average of the known data values. Hence, if there are n known values (Z_1 to Z_n), the value (Z) of the parameter at some intermediate location is determined by:

$$Z = \sum_{i=1}^{n}(Z_i/d_i)\Big/\sum_{i=1}^{n}(1/d_i)$$

(10.1.2)

3. **Polynomial fit.** Rather than taking just the inverse of the distance, a polynomial function may be used. The coefficients of this polynomial function may be determined from the data itself by finding the set of coefficients that fits the data best for the known well locations.
4. **Kriging.** Kriging is an advanced technique that involves using all the data available to determine the best combination of the available data points at a particular intermediate location. In order to do this, it is first necessary to mathematically describe how the parameter in question varies between the known data points. This is done by constructing a semivariogram of the data. A semivariogram may be constructed using the formula:

$$\text{gamma}(h) = (1/2 * N(h)) * \sum_{i=1}^{i=N(h)} (V_{xi',yi'} - V_{xi,yi})^{\wedge}2 \qquad (10.1.3)$$

where $V_{xi',yi'}$ = (known) value of the parameter at point xi', yi'
$V_{xi,yi}$ = (known) value of the parameter at point xi, yi
h = distance from xi, yi to xi', yi'
$N(h)$ = the number of pairs that are a distance h apart.

A plot of gamma(h) vs. h constitutes the semivariogram. The semivariogram captures the contribution to the total uncertainty when using a data point that is a certain distance away from the point at which the estimation is being made. Kriging involves finding a set of weighting factors (done automatically within a computer) that minimizes the total uncertainty in the estimate made at the intermediate point. Moreover, kriging also provides the variance of the estimation error from which an error map can be drawn.

10.1.4 Quantitative Analysis from Maps

While maps are invaluable in helping the petrophysicist and geologist understand the areal variation of properties, they may also be used for quantitative analyses. In most petroleum engineering departments, quantitative work is always done on the computer these days. However, it is always recommended to make a reality check using more basic techniques, since lack of full understanding of how software works, or bugs in the program, can lead to erroneous results.

Initially we will show how GBV may be determined from maps, then extend the concept to show how HIIP can also be determined. Consider an oil reservoir where both the top and base of the structure have been mapped, and the oil/water contact (OWC) is known. The first step is to make an area-depth map for the top of the structure. This is done by measuring the area contained within each contour, starting at the shallowest and working gradually deeper, until one is at the first contour that falls below the OWC.

The area contained within this contour, since it is an irregular shape, is most often measured using a device called a planimeter, which (once the area has been completely traced) will determine the area contained therein. Obviously, since the planimeter measures only the area on the paper, a conversion has to be made using the map scale. Hence, for example, if the map is 1:25,000, a traced area of X square centimeters would be equivalent to a real area of $(25000)^2*X/(100)^2$ square meters. Once this has been done, a plot is made on graph paper of the area (in square metres) against the depth (Figure 10.1.2).

The same is now done for the base of the structure, and the two measurements are combined. A line is drawn indicating the position of the OWC (Figure 10.1.3).

By planimetering (or just counting squares on the graph paper), one can determine the area (in cm^2) on the area-depth graph representing the

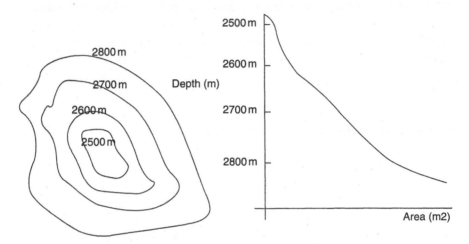

Figure 10.1.2 Maps and Area-Depth Graphs

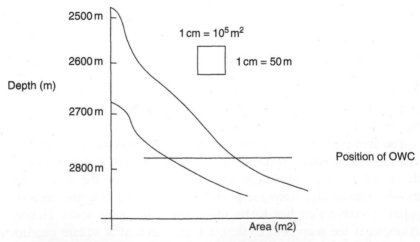

Figure 10.1.3 Gross Bulk Volume (GBV) from Area-Depth Graph

volume of the GBV of the oil column. To convert this area into a volume: Draw a square of side 1 cm on the graph paper. Along one horizontal side, indicate how many square meters are represented by 1 cm. Likewise, on the vertical side indicate how many meters in depth are represented by 1 cm. Taking the product of these two conversion factors will yield the conversion factor to convert square centimeters into cubic meters of rock.

In the example shown in Figure 10.1.3, if the area measured from the graph were $10 \, cm^2$, the GBV of the oil column would be given by:

$$GBV = 10 * 10^5 * 50 = 5 * 10^7 \, m^3.$$

The technique illustrates well why there are problems with using net/gross when reservoir quality varies in depth in a reservoir. This can be seen as follows. Imagine that the horizon shown in the example consists of a very good quality sand overlying a nonproducible shale. Say the porosity in the sand is 30% and $S_w = 0.10$. However, taking the whole interval into consideration, the net/gross is only 30%. One would estimate the STOIIP (stock tank oil initially in place) to be:

$$STOIIP = GBV * (0.3) * (0.9) * (0.30)/B_o$$

where B_o is the oil formation volume factor. Now consider what would have happened if the base of the good sand were mapped instead of the entire package. A new GBV (GBV′) would be determined by planimetering the new area-depth graph. The net/gross corresponding to GBV′ is now 1.0 instead of 0.30, but of course GBV′ is less than GBV. Another estimate of the STOIIP is then given by:

$$STOIIP' = GBV' * (0.3) * (0.9) * (1.0)/B_o.$$

Because of the nature of the way that the OWC cuts across the structure, it is certainly not the case that STOIIP = STOIIP′. In essence, because the poor-quality rock falls preferentially below the OWC, one is underestimating the STOIIP by using the coarser interval and applying a net/gross. Conversely, if the good-quality rock had been located at the base of the sequence instead of at the top, the STOIIP would have been overestimated through using net/gross.

The concept of area-depth mapping may also be applied to other parameters. For instance, if a map is made of the EHC (net*porosity*hydro-

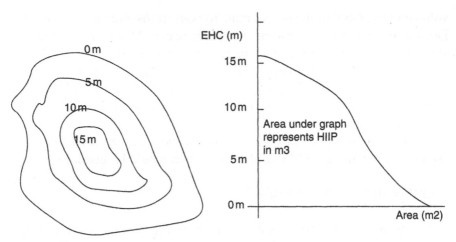

Figure 10.1.4 Hydrogen Initially in Place (HIIP) from Area-Depth Graph

Table 10.1.1
Example of well data

Well	Easting, km	Northing, km	Depth of Top Sand, m tvdss*
1	100	100	3000
2	100.5	100.7	3100
3	101	100	3200
4	100.6	99.3	3100
5	100	99.2	3080
6	100	101.5	3120
7	99.3	99.3	3110
8	98.5	100	3220
9	99.5	100.6	3090

*m tvdss = meters of true vertical depth subsea

carbon saturation), this may also be planimetered to yield the STOIIP directly, as is illustrated in Figure 10.1.4.

Exercise 10.2. Area-Depth Graph

An oil accumulation is discovered by wells penetrating the structure at the locations/depths shown in Table 10.1.1.

The thickness of the sand is uniformly 325 m, with porosity 20% and average S_w of 20%. The OWC was found at 3,070 m at TVDss (total vertical depth subsea). B_o is 1.3 rb/stb (reservoir barrels/stock tank barrels)

1. Make a map of the top structure and draw contours every 25 m.
2. Construct area-depth maps for the top and base of the structure.
3. Estimate the GBV, NPV, and STOIIP.
4. Consider what the effect on STOIIP would be if in fact S_w were 10% for depths shallower than 25 m above the contact, and 30% for 0–25 m above the contact.

10.2 BASIC GEOLOGICAL CONCEPTS

10.2.1 Clastic Reservoirs

Clastic rocks are defined as being composed of consolidated sediments formed by the accumulation of fragments derived from preexisting rocks and transported as separate particles to their places of deposition by purely mechanical agents. These fragments may be transported by water, wind, ice, or gravity. The manner of their movement may be by suspension, saltation, rolling, or solution. The effect of the transportation is to change any of the following characteristics of the fragments (or grains):

- Size
- Shape
- Roundness
- Surface texture
- Orientation
- Mineralogical composition

Measurement of these changes may provide information on the transportation mechanism. The types of environmental deposition are as follows:

Desert
Waddis
Aeolian dune systems
Sabkhas
Fluvial
Alluvial fans
Floodplains

Channel systems
Deltaic
Distributary channels
Delta plain
Delta front
Glacial
Loess
Tillites
Varves
Deep marine
Turbidite fans
Slumps, slides, debris flows
Turbidity currents
Shoreline
Beaches, bars, barrier islands
Coastal aeolian plains
Cheniers
Swamps, marshes, estuaries

Knowledge of the environment of deposition is crucial to understanding how the good-quality sands, if any, are likely to be distributed over a prospect. Mineral composition of clastic reservoirs, in order of abundance are: quartz, clay minerals, rock fragments, feldspars, chert, mica, and carbonate fragment.

A "clean" sandstone will comprise mainly quartz grains. This may have a porosity as high as about 40% and permeability up to 5 darcies. The presence of clays and minerals forming cement between the grains will have the effect of reducing the porosity and permeability, as will different grain distributions.

The distribution of clay minerals is also related to the environment of deposition and is of particular significance to the petrophysicist, in view of the effect that clay has on permeability, conductivity, and water saturation. Clays found in sandstones are classed as either allogenic or authigenic. Allogenic clays are those that were present prior to deposition. These may take the following forms:

- Individual clay particles, dispersed as matrix or as laminae
- Pellets formed from clay flocculation or excreted by organisms
- Clay aggregates derived from preexisting shales outside the depositional basin

Authigenic clays are those that have formed at the same time or after deposition. Three processes are possible that lead to their presence:

- Allogenic clays may be transformed into new types through the effect of temperature, pressure, and pH.
- Clay mineral may be formed by the diagenesis of nonclay minerals such as feldspars, pyroxenes, amphiboles, and micas.
- Clay mineral may be precipitated from porefluids.

Principal clay minerals are as follows:

- KAOLINITE, $Al_2Si_2O_5(OH)_4$. The clay is usually in the form of hexagonal crystals, which may be stacked to form accordion-type shapes. These structures may fill the pores and have an impact on permeability. However, the cation exchange capacity (CEC) of kaolinite is low, meaning that it absorbs relatively little water.
- CHLORITE, $(Mg, Fe)6\ AlSi_3O_{10}(OH)_8$. Chlorite may form in many different shapes, such as plates, rosettes, honeycombs, or round growths. Typically it will coat the sand grains and pore throats, having a negative effect on permeability. However, unlike some other clays, it does react to acid, so permeability around the wellbore may be increased via acid stimulation. CEC is also relatively low.
- ILLITE, $(H_3O, K)\ (Al_4Fe_4Mg_4Mg_6)(Si_7Al)O_{22}\ (OH)_4$. Illite can be particularly damaging to permeability when it takes the form of hairlike structures that may block the pore throats. It does react partially to strong acid. The CEC is higher than that of chlorite or kaolinite but lower than that of montmorillonite.
- MONTMORILLONITE, $(Na, K, Mg, Ca)\ Al_2Si_4O_{10}(OH)_2H_2O$. Montmorillonite has the highest CEC and has therefore the greatest effect on saturation through clay-bound water. Swelling of the clay can also cause problems when drilling with freshwater mud. The clay takes the form of crinkly coatings on detrital grains, or a cellular structure similar to honeycomb chlorite.

10.2.2 Carbonate Reservoirs

Carbonates originate from the calcareous skeletons of organisms, forming bioclastic sediments. These fragments are cemented by carbonate precipitating from water. Most of the organisms lived on the bottom in shallow marine water, where algae were present. However, after dying, the organisms may have fallen to a greater depth and accumulated. Below a certain depth (4,000–6,000 m), all carbonate is dissolved as a result of the high pressure.

The main difference between carbonate and clastic reservoirs is that clastic deposition requires the transportation of grains to the sedimentary basin, whereas carbonates originate within the basin of deposition. Since the effect of clastic deposition is to typically cloud the water, making the environment unsuitable for organisms relying on photosynthesis, it is usually not possible to have carbonate and clastic reservoirs coexisting. However, it is of course possible for one to be overlying another due to changes in the environment of deposition over geological time.

Carbonate reservoirs comprise the following types:

- *Shallow marine carbonates.* The rate of skeletal production in shallow marine water is generally high. These skeletons break down, due to action by crustaceans and fish or by turbulence. The effect is to generate carbonate sediment that may be transported to the final place of deposition. This sediment may be modified by burrowing organisms. Fecal pellets so generated may form grains, and hence result in porosity.
- *Deepwater carbonates.* Deepwater carbonates are deposited at a depth below that at which photosynthesis occurs. Typically the sediments are formed from oozes consisting of skeletons of pelagic organisms.
- *Reefs.* Reefs are built by calcium carbonate–secreting organisms growing on the remains of previous generations. The large skeletal organisms (e.g., corals) generally remain in place after death, and this may result in the formation of cavities partially filled with sediment. Most reef sediment is produced by segmented (e.g., crinoids, algae) or nonsegmented organisms (bivalves, brachiopods, foraminifera) that grow in the spaces left by larger skeletal organisms.

Initially, the porosity in calcisands (i.e., matrices comprising carbonate grains) is very high (45% porosity and 30 D permeability). However, postdepositional diagenetic processes have the effect of drastically reducing this. Factors that reduce the porosity are:

- Cementation: precipitation of $CaCO_3$ from the pore waters into the porespace
- Internal sedimentation: or filling of the porespace by sediment
- Compaction: grain repacking
- Pressure solution: dissolution of $CaCO_3$ in one part and precipitation in the porespace of another part

Processes that may increase the porosity are:

- Leaching: This can be fabric related or nonfabric related. Fabric-related leaching is selective due to mineralogical differences in the sediment and results in mouldic porosity. The nonfabric-related type tends to lead to large voids or karsts.
- Dolomitization: This is the replacement of $CaCO_3$ by $CaMg(CO_3)^2$. Where this creates porosity, it will generally be vuggy or intercrystalline in nature.

10.2.3 Faulting and Deformation

In any field, the forces acting on the sediments will be both gravitational and either extensional or compressional in any horizontal direction. Extensional features may result from (a) lack of lateral support nearby sediments, (b) movement of basement rocks, or (c) instabilities in the overburden arising from differential compaction or salt diapirs. Compressional features may result from (a) gravity-sliding of rocks over inclined basement surfaces, or (b) movement of basement blocks.

The main features observed in extensional tectonics are as follows:

- *Growth faults.* These are faults in which the thickness of rocks in the downthrown block are greater than those (in the same time units) in the upthrown block. The downthrown block has subsided quicker than the upthrown block and acquired more sediment. The fault plane is usually curved upward. This is illustrated in Figure 10.2.1.
- *Rollover anticlines.* These are growth anticlines in which rock units thicken from the crest toward the flanks. The flanks subsided faster than the crest and accumulated a greater thickness of sediment.
- *Normal faults.* Movement in the basement either during or after deposition may cause faulting. In normal faulting the sediments on the upper side of the fault plane move vertically downward relative to those on the lower side of the fault plane. This is illustrated in Figure 10.2.1.
- *Salt domes.* During Permian and Triassic times, large volumes of seawater became isolated, and subsequent evaporation created thick salt deposits. Since the salt is essentially ductile where it is confined, there is a tendency for it to bulge at the weakest point through the overlying rock, forming a diapir. This diapir will cause radial fractures in the overlying rock around it. The diapir may itself form part of the seal, allowing hydrocarbons in sands to be trapped.

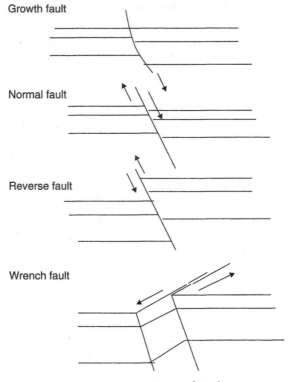

Growth fault

Normal fault

Reverse fault

Wrench fault

Figure 10.2.1 Types of Fault

Features associated with compressional tectonics are:

- *Reverse (or thrust) faults.* Where the main stress in the reservoir is horizontal on one axis, with the second main stress also horizontal but normal to this (i.e., the stress in the vertical direction is the weakest), reverse faulting may occur. The frictional forces associated with rocks passing over one another are very high, and overpressures may often be associated with reverse faulting, since they provide a mechanism whereby the friction may be reduced (see Figure 10.2.1).
- *Wrench faults.* These will occur where the second highest stress is in the vertical direction and the weakest in the horizontal direction normal to the axis of greatest stress. This results in a horizontal displacement of one block relative to the other.
- *Folding.* Compressional stresses will generally give rise to faulting in brittle rock. In more ductile rock, folding is possible, and this can also

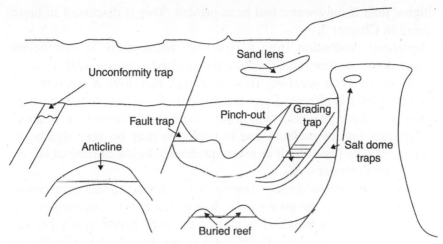

Figure 10.2.2 Types of Trap

provide a sealing mechanism. Terminology useful to folding is as follows:

- Anticline: convex up
- Syncline: concave up
- Hinge line: the line of maximum curvature
- Limb: area of least curvature or flank
- Crest: topographically highest point
- Trough: topographically lowest point
- Axial plane: the plane defined by hinge lines of all horizons
- Crestal plane: the plane defined by crests on all horizons

The basic types of hydrocarbon trap are illustrated in Figure 10.2.2.

10.2.4 Abnormal Pressures

Normal pressures are defined by taking a hydrostatic gradient (typically 0.45 psi/ft or 1.04 g/cc) from surface to a certain depth. Significant deviation from the pressure so determined would be termed abnormal. The principal mechanisms giving rise to abnormal pressures are:

- *Low water table or high elevation.* In mountainous areas, the water table may lie some depth below surface; and in the upper part, any hole drilled will be through dry rock.
- *Hydrocarbons.* The presence of a hydrocarbon column above an aquifer will cause the pressure at a depth in the hydrocarbon column to be

higher than if only water had been present. This is discussed in more detail in Chapter 2.

- *Depletion.* Production from a reservoir, where there is insufficient water drive or injection to replace the drained volumes, will result in a loss of reservoir pressure. To some extent, this drop in pressure will be offset by the effects of gas coming out of solution or compaction. However, these processes will always lag behind production. In some reservoirs, loss of pressure due to depletion may be very significant (up to 5000 psi) and result in severe problems during drilling of subsequent development wells.

- *Compaction disequilibrium.* During burial under equilibrium conditions, the water in the porespace is free to leave, thus ensuring that as the overburden increases, it is mainly the rock matrix that takes the weight of the overburden. If the water is not, or only partially, free to leave, some of the weight of the overburden is taken up by the porefluid, resulting in a much higher porefluid pressure than would otherwise be the case. Although shales are usually impermeable over production life cycles, the long time periods associated with deposition and burial usually cause them to expel water quickly enough for equilibrium to be maintained. Conditions in which such overpressures are likely to occur are when (a) the permeability of the shale decreases with compaction, (b) the thickness of the shale is very great, (c) the shale is structurally weak, or (d) the rate of burial is very fast.

- *Aquathermal pressures.* Where part of a system becomes isolated so that it retains a constant volume under burial, a change in temperature may result in a rapid increase of pressure.

- *Phase changes.* The volume of water in a system may increase, thereby resulting in an increase of pore pressure under conditions of (a) dehydration and dewatering of clays, in particular where montmorillonite is transformed into illite, and (b) conversion of gypsum to anhydrite.

- *Osmosis.* Where two reservoirs have different salinity and are separated by a semipermeable membrane (e.g., a clay), water will flow from the less saline to the more saline, resulting in an increase in pressure. In theory, such pressure differentials could reach 3000 psi, although this has not been observed in practice.

RESERVOIR ENGINEERING ISSUES

11.1 BEHAVIOR OF GASES

Consider a gas for which the behavior is as shown in Figure 11.1.1.

The line dividing the regions where the substance is liquid or gas is called the **vapor pressure line**. Hence, crossing the line from left to right, the liquid boils and becomes a gas; and in crossing from right to left, the gas condenses to form a liquid. Above the critical point (defined by P_c and T_c), there exists only a fluid phase with gas indistinguishable from liquid.

An equation of state relates the pressure, volume, and temperature of a substance. For gases at low pressures and medium to high temperatures, the ideal gas law may be applied:

$$pV = nRT \tag{11.1.1}$$

where
p = pressure in N/m^2
V = volume in m^3
n = number of moles
R = gas constant (8.3143 joules/Kelvin/mole or 10.732 psia.cu ft/lb mole deg R)
T = temperature in Kelvin (°C + 273).

This equation is useful for quickly estimating how the volume of a gas might change when taken from conditions of one pressure/temperature to another. For example, say there was an influx of $1\,m^3$ of gas into a borehole, where the pressure was 3000 psi ($4.35 * 10^6\,N/m^2$) and 150°C (423 K). The volume at 2500 psi/130°C could be estimated as follows:

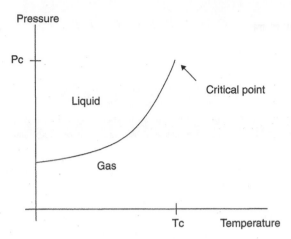

Figure 11.1.1 P-T Diagram

$$nR = \text{constant} = P_1V_1/T_1 = P_2V_2/T_2. \tag{11.1.2}$$

Hence, $V_2 = V_1 * (P_1 * T_2 / (P_2 * T_1))$.

In the example, $V_1 = 3000$, $T_1 = 423$, $P_2 = 2500$, $T_2 = 403\,\text{K}$. Hence, $V_2 = (3000 * 403 / (2500 * 423) = 1.14\,\text{m}^3$. A more correct equation for real gases as encountered in reservoirs is that of van der Waals:

$$(p + a/V^2) * (V - b) = RT. \tag{11.1.3}$$

This equation may be expressed in the form:

$$pV = nZRT \tag{11.1.4}$$

where Z varies with pressure and temperature. In order to calculate Z, it is first necessary to know, for the gas concerned, the values of the critical pressure and temperature defined by P_c and T_c. These are used to determine the reduced pressure and temperature given by:

$$P_r = P/P_c$$

$$T_r = P/P_c. \tag{11.1.5}$$

The most commonly used experimentally determined correlation between Z and P_r/T_r is that of Standing and Katz, shown in Figure 11.12.

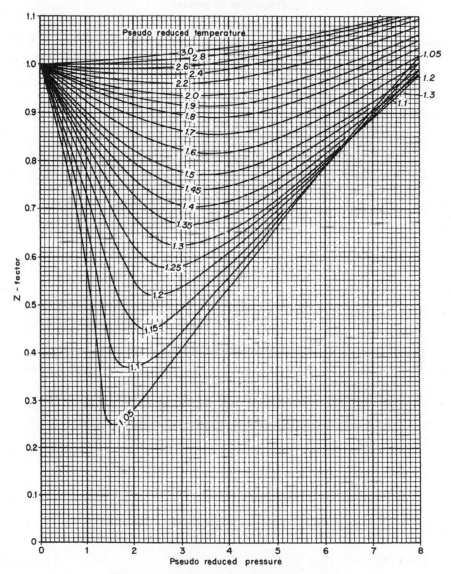

Figure 11.1.2 Standing and Katz Correlation

The critical pressure and temperature may be measured experimentally. Some values of common gases are as follows:

P_c and T_c may also be estimated from the formulae:

$$P_c = 4.55 * 10^6 + 0.345 * 10^6 * \gamma - 0.383 * 10^6 \gamma^2 \qquad (11.1.6)$$

Table 11.1.1
Properties of gasses

Gas	Formula	Molecular Mass (kg/kmol)	P_c (N/m²)	T_c (K)
Methane	CH_4	16	$4.61*10^6$	191
Ethane	C_2H_6	30	$4.88*10^6$	306
Propane	C_3H_8	44	$4.25*10^6$	370
Nitrogen	N_2	28	$3.4*10^6$	126

$$T_c = 100 + 167 * \gamma \tag{11.1.7}$$

where γ is the specific gravity of the gas, defined as:

$$\gamma = \text{(density of gas at standard conditions)}/ $$
$$\text{(density of air at standard conditions)}. \tag{11.1.8}$$

Exercise 11.1. Density of Air

Calculate the density of air at standard conditions: temperature = 15.5°C, pressure = 1 atm. The molecular mass of air should be taken as 29 kg/kmol.

The gas expansion factor (E) relates the volume of a fixed number of moles of gas at standard conditions relative to reservoir conditions. It is defined as:

$$E = \text{volume at standard conditions } (15.5°C, 1.10325*10^5\ N/m^2)/$$
$$\text{volume at reservoir conditions.} \tag{11.1.9}$$

Using equation 11.1.4, it can be shown that:

$$E = 1.965*10^{-4}\ P/(Z*T) \tag{11.1.10}$$

where P is in N/m² and T is in K. Very often you will see this equation in oilfield units, in which case it becomes:

$$E = 35.37 * P/(Z*T) \tag{11.1.11}$$

where P is in bar and T is in Rankine (°R = °F + 460 = °K * 1.8).

11.2 BEHAVIOR OF OIL/WET GAS RESERVOIRS

Oil reservoirs are generally multicomponent systems. In other words, there is gas present, either in solution or as free gas in the reservoir. Hence, a single-phase diagram such as that shown in Figure 11.1.1 is not appropriate. In systems that consist of more than one component, the points of phase coexistence constitute a two-dimensional region bounded by the dewpoint line and bubble point line. This is illustrated in Figure 11.2.1.

Above the bubble point (in terms of pressure at a certain temperature), only liquid may exist, i.e., any gas is completely dissolved in the oil. Above the dewpoint (in terms of temperature at a certain pressure, all the hydrocarbons are gaseous.

Reservoirs will typically fall into one of three categories, as shown in Figure 11.2.2.

In wet gas reservoirs, the initial reservoir pressure and temperature lie to the right of the dewpoint line. Within the reservoir during production, as the pressure drops, the dewpoint is not crossed and the hydrocarbons remain gaseous. However, at surface, where the temperature is allowed to drop to ambient conditions, the dewpoint will be crossed and condensate liquid will drop out of the gas.

In retrograde condensate reservoirs, the initial pressure and temperature are above the dewpoint, but during production the dewpoint is crossed in the reservoir. This leads to liquid forming in the reservoir (retrograde

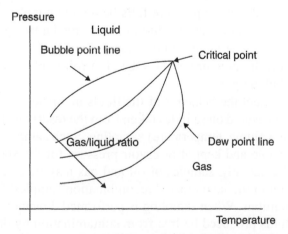

Figure 11.2.1 P-T Diagram for a Two-Phase System

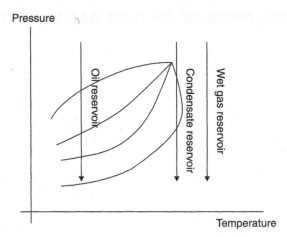

Figure 11.2.2 Types of Reservoirs

condensation). Later on, the pressure may fall so that the dewpoint is crossed again and the condensate may evaporate in the reservoir. At surface a combination of liquid and gas will be produced.

In oil reservoirs the initial pressure is such that all the gas is dissolved in the liquid at reservoir conditions. During depletion, the reservoir pressure may fall below the bubble point and gas may start to come out of solution in the reservoir. Even if the reservoir pressure is kept above the bubble point, when the fluids reach surface, the gas will have come out of solution to be produced as associated gas. In many structures, the pressure near the OWC (oil/water contact) is above the bubble point, but at a certain shallower depth the pressure falls below the bubble point. At this point gas will have come out of solution and formed a free gas cap occupying the crest of the structure. In this case, at least under virgin equilibrium conditions, the pressure at the GOC (gas/oil contact) is equivalent to that of the bubble point.

In order to predict the behavior of the fluids in a reservoir, the properties should be measured on samples taken from the reservoir. This is called PVT (pressure, volume, temperature) sampling. Such samples will ideally be taken downhole and kept at reservoir pressure for transporting to the laboratory. Alternatively, samples of oil and gas may be taken at surface and recombined in the laboratory ("recombination samples") to re-create downhole conditions. When sampling is performed downhole, it is essential that the fluids produced be free from contamination by drilling fluids and that the drawdown is such that samples are taken above the bubble point.

A fundamental problem with PVT sampling and measurement is the fact that the end point of a process whereby the gas is liberated from oil is dependent on whether the gas is kept in close proximity with the oil during the liberation or is removed. In a reservoir, the gas typically remains in contact with the oil and will follow the pattern of behavior as indicated by the pressure-temperature (P-T) diagram. However, during production in a well, the gas will become isolated from the oil. This isolation changes the phase behavior of the oil left behind, modifying the P-T diagram. Hence, unless any experiment models the true processes undergone by the gas and oil during migration to the wellbore, production, and gas/oil separation at surface, it cannot accurately predict the actual production of gas and oil from a field.

The parameters that are measured during PVT analysis consist of the following:

- **The oil formation volume factor** (B_o). B_o is defined as the volume of reservoir liquid required to produce one volume unit of stock tank oil (i.e., oil at standard surface conditions). Depending on how "gassy" the oil is (i.e., how high the gas/oil ratio [GOR] is), B_o typically varies between 1.0 and 1.5.
- **The gas formation volume factor** (B_g). B_g is the reservoir volume of one volume unit of gas at standard conditions. Rather confusingly, while in SI units B_g is normally given in reservoir m^3/standard m^3, in field units it is usually quoted as reservoir barrels/standard cubic feet. Typical range (for SI units) is 0.004–0.06 r.m^3/st.m^3.
- **The solution gas/oil ratio** (R_s). R_s is defined as the volume units of gas that evolve from B_o reservoir volume units of oil when the oil is transported to surface conditions. It has the units of standard m^3 gas/standard m^3 oil, or standard cubic feet/standard stock tank barrel (stb). Where there is no free gas being produced from the reservoir, this is the same as the GOR. However, GOR is a term that is used to describe well production behavior, unlike R_s, which is essentially a laboratory measurement.

If the reservoir pressure falls below the bubble point, free gas will be produced in the reservoir. Since the gas is more mobile than the oil, this will lead to a dramatic rise in the gas produced at surface, although in time the GOR will drop back once most of the free gas has been produced. Some of the free gas may move to the crest of the structure and form a secondary gas cap.

Where the producing GOR in wells has become high, the oil rate can be improved by injecting water into the reservoir. This will have the effect of raising reservoir pressure, causing some free gas to redissolve, and also help displace oil to the wells. However, where water break-through occurs in the producing wells, the overall effect on production may be negative.

11.3 MATERIAL BALANCE

Material balance treats a reservoir like a tank, with areal and depth pressure equilibrium. For undersaturated oil reservoirs (i.e., those for which the pressure remains above the bubble point), oil is produced by the expansion of the liquid phases and connate water, the shrinkage of the pore volume, and the influx of water from the aquifer. When the reservoir pressure drops from its initial value P_i to P, the formation volume oil factor increases from B_{oi} to B_o.

If the produced oil volume at standard conditions is N_p (with initial volume N), the volume removed from the reservoir at downhole conditions must be $N_p * B_o$. The expansion of the oil in place must be $N * B_{oi} - N * B_o$. The initial pore volume equals $N * B_{oi}/(1 - S_w)$, where S_w is the water saturation. If the compressibility of the water is C_w, and that of the pores is $C\phi$, the composite compressibility C is given by:

$$C = (C_w * S_{wc} + C\phi)/(1 - S_w). \qquad (11.3.1)$$

From material balance it must be true that produced volume (at reservoir conditions) = expansion of oil in reservoir + expansion of pore volume + water influx (W). Hence:

$$N_p * B_o = N * (B_{oi} - B_o) + N * B_{oi} * C * (P_i - P) + W. \qquad (11.3.2)$$

Porespace compressibility can be measured in the lab as part of a special core analysis (SCAL) program. Typical values range from about $15 * 10^{-5}$ l/bar for a low-porosity rock to $5 * 10^{-5}$ l/bar for a high-porosity rock. Water compressibility varies with pressure, temperature, salinity, and amount of dissolved gas. An approximate value to use is $4.35 * 10^{-5}$ l/bar.

Exercise 11.2. Material Balance of Undersaturated Oil Reservoir

Consider a reservoir with negligible water drive ($W = 0$). Bubble point is 200 bar.
If $N = 10^7$ standard m³,
$B_{oi} = 1.3$,
B_o at 250 bar = 1.25,
$S_w = 0.2$,
$C\phi = 10.0 * 10^{-5}$ l/bar,
$C_w = 4.4 * 10^{-5}$ l/bar, and
$P_i = 300$ bar,
how much oil will have been produced when the reservoir pressure has dropped to 250 bar?

11.4 DARCY'S LAW

For linear flow, an empirical equation developed by Darcy is as follows:

$$u = -(k/\mu) * (\partial p/\partial x + \rho * g * \partial z/\partial x) \qquad (11.4.1)$$

where:
u = flow velocity in m/s
k = permeability in m² (note that 1 darcy is 10^{-12} m²)
μ = viscosity, in Pa (note that $1\,cp = 10^{-3}$ Pa)
p = pressure of flowing phase, in Pa
z = vertical distance upward, in m
x = distance in the direction of flow, in m
g = gravitational constant (9.81), in m/s²
ρ = density of the flowing phase.

For purely horizontal flow over a distance L (having pressure drop ΔP) with area A, this equation reduces to:

$$Q = -(k/\mu) * \Delta P * A/L \qquad (11.4.2)$$

where Q is flow rate measured in m³/s. The above equation also approximately applies in so-called Darcy units, where Q is in cc/sec, k is in darcies, A is in cm², μ is in cp, P is in atmospheres, and L is in cm. The approximation originates from the fact that 1 atm is in fact 1.01325 * 10^5 Pa, and not 10^5 Pa.

In oilfield units, and incorporating the conversion from downhole to surface conditions via B_o, the equation becomes:

$$Q = -1.127 * 10^{-3} * (k/\mu) * \Delta P * (B_o * L)$$

where Q is in stb/day, k is in md, A is in ft^2, μ is in cp, P is in psi, and L is in ft.

For radial flow into a borehole, incorporating the porosity and effect of fluid/pore compressibility, Darcy's equation becomes:

$$(1/r) * \partial/\partial r (r * \partial p/\partial r) = \partial p/\partial t * \phi * C * k/\mu \qquad (11.4.3)$$

where r = radial distance from the center of the borehole (in m), ϕ = porosity, and C = composite compressibility, in 1/Pa. The composite compressibility is given by:

$$C = C_o * (1 - S_w) + C_w * S_w + C_\phi \qquad (11.4.4)$$

where compressibility is expressed as C_o for oil, C_w for water, and C_ϕ for porespace.

A system is said to be in a steady state when pressure does not vary as a function of time, i.e., $\partial p/\partial t$ is zero. A system is said to be in a semi-steady state when $\partial p/\partial t$ is a constant.

Every closed finite system that is produced at a constant rate will asymptotically approach a semi-steady state. An infinitely extended system will never approach a semi-steady state.

Equation 11.4.3 has solutions depending on the boundary conditions that are applied. For a situation in which the boundary of the area being drained is maintained at a constant pressure, a steady state will eventually develop for which:

$$P - P_w = (Q * \mu/(2 * \pi * k * h)) * (\ln(r_e/r_w) - 1/2) \qquad (11.4.5)$$

where:

P = mean pressure in the drainage area, in Pa
P_w = bottomhole flowing pressure, in Pa
Q = flow rate in reservoir, m^3/sec
h = thickness of reservoir, in m

r_e = drainage radius, in m
r_w = wellbore radius
k = permeability, in m^2
μ = viscosity, in Pa.

For a situation in which the boundary is sealed and the pressure is dropping linearly with time, a semi-steady state exists that has the solution:

$$P - P_w = (Q*\mu/(2*\pi*k*h))*(\ln(r_e/r_w) - 3/4). \tag{11.4.6}$$

Note that P and P_w both vary with time, but their difference remains constant. Since the wellbore radius is always very small compared with the drainage radius, it is possible mathematically to study the theoretical case in which the wellbore radius becomes infinitessimally small, whose solution is:

$$P_i - P_f = Q*\mu/(4*\pi*k*h)*E_i(x) \tag{11.4.7}$$

where P_f is the pressure of the flowing phase, $E_i(x)$ denotes the exponential integral, and

$$x = \kappa*r^2/4*t$$

$$\kappa = \phi*\mu*C/k.$$

The exponential integral, $E_i(x)$, is given by:

$$E_i(x) = -\gamma - \ln(x) - \Sigma(-x)^n/n!*n$$

where γ is Euler's constant (1.781). If $t > 25*\phi*\mu*C*r^2/k$, the summation term becomes less than 0.01 and can be ignored; then we can give the approximate version of equation 11.4.7 as:

$$P_i - P_f = -(Q*\mu/(4*\pi*k*h)*\ln(\gamma*\kappa*r^2/4*t). \tag{11.4.8}$$

Solving for t/r^2:

$$t/r^2 = (1/4)*\gamma*\kappa*\exp(4*\pi*k*h*\Delta P/(Q*\mu) \tag{11.4.9}$$

where $\Delta P = P_i - P_f$.

Exercise 11.3. Radial Flow

Calculate the time taken for the flowing pressure to fall by 30 bar in the vicinity of a well subject to the following flowing conditions: $k = 300\,\text{md}$, $h = 30\,\text{m}$, $R_w = 15\,\text{cm}$, $\phi = 0.20$, $\mu = 10\,\text{cp}$, $C = 2*10^{-4}/\text{bar}$, $Q = 200\,\text{res.m}^3/\text{day}$,

11.5 WELL TESTING

The equations given in the previous section can be used to predict flow performance in ideal cases. However, in reality, flow near the wellbore is influenced by formation damage caused by drilling and production processes (precipitation, fines moving, gas blocking, water dropout, etc.). The pressure difference between the bottomhole well flowing pressure and the reservoir pressure, ΔP, relates to formation properties via a formula of the form:

$$\Delta P = Q * \mu * J / (2 * \pi * k * h) \tag{11.5.1}$$

where J may be considered a dimensionless productivity. In the case of steady state flow, $J = \ln(r_e/r_w) - 1/2$; and for semi-steady state flow, it is $\ln(r_e/r_w) - 3/4$. The productivity index (PI) is defined as:

$$PI = Q/\Delta P. \tag{11.5.2}$$

The effect of formation damage is that the PI is smaller than one would expect from theoretical values, or ΔP is greater than one would expect for a given Q. This damage may be quantified by introducing a term called S such that:

$$\Delta P = Q * \mu * (J + S) / (2 * \pi * k * h). \tag{11.5.3}$$

If the radius of the damaged zone is r_1, and it has permeability k_1, it may be shown theoretically that S is given by:

$$S = (k/k_1 - 1) * \ln(r_1/r_w). \tag{11.5.4}$$

This is subject to the following assumptions:

- Flow is radially symmetric.
- Porosity and compressibility are not affected by the damage.
- Pressure is continuous over the boundary between the damaged and virgin zones.
- The damaged zone is in the steady state and the region outside is in either the steady or semi-steady state.

The effect of stimulation through acid, steam, reperforation, etc., will be to reduce S. Of particular interest to the petrophysicist is how the product $k*h$ is derived from a well test, since this will have to be reconciled with values derived from logs. During the initial phase of testing a well, the semi-steady state has not been reached, and a transient state will prevail. In this regime the following equations apply:

$$2*\pi*k*h*\Delta P/(Q*\mu) = (1/2)*\ln(4*t_d/\gamma) + S \qquad (11.5.5)$$

where t_d is the dimensionless time given by:

$$t_d = k*t/(\phi*\mu*C*R_w^2). \qquad (11.5.6)$$

If two measurements are made of P and t, kh can be derived from:

$$kh = (1/2)*Q*\mu*\ln(t_1/t_2)/(2*\pi*(P_2 - P_1)). \qquad (11.5.7)$$

Having determined kh, S may be determined from equation 11.25. Note that in the equations presented, SI units must be used throughout, and Q is in reservoir volume, not standard volume.

In reservoir limit testing, a well is produced sufficiently long for the semi-steady state to be reached. Once semi-steady state is reached, the well pressure will decline linearly with time, and the rate of change of pressure with time will be constant. In the semi-steady state:

$$Q = -C*A*h*\phi*(dP/dt). \qquad (11.5.8)$$

Hence if measurements of P vs. t are made, the drainage area A can be deduced. Most commonly, wells are tested by letting them flow for a certain period of time, then shutting them in and observing the buildup of pressure with time. If the well is flowed for a period T, then shut in, the pressure behavior will simply follow that of the pressure observed in a well flowed from $t = T$ subtracted from that of a well flowed from $t = 0$. Note that the pressure observed thereby becomes independent of S, which cancels out.

When analyzing buildups, it is common, in addition to defining the dimensionless time, to introduce the dimensionless pressure P_d given by:

$$P_d = 2 * \pi * k * h * (P_i - P)/(Q * \mu). \tag{11.5.9}$$

Then it can be shown that:

$$2 * \pi * k * h * (P_i - P_w)/(Q * \mu) = P_d(T_d + \Delta t_d) - P_d(\Delta t_d). \tag{11.5.10}$$

In this equation, $P_d(T_d + \Delta t_d)$ represents the pressure at a dimensionless time $T_d + \Delta t_d$, converted into a dimensionless pressure using equation 11.5.9. For an infinite reservoir:

$$P_d = (1/2) * \ln(4 * t_d / \gamma). \tag{11.5.11}$$

Hence

$$2 * \pi * k * h * (P_i - P_w)/(Q * \mu) = (1/2) * \ln[(T_d + \Delta t_d)/\Delta t_d]. \tag{11.5.12}$$

This is the basis of the so-called Horner plot.

If pressure measurements are made at two different times (P_1 and P_2), it can be shown that:

$$k * h = (1/2) * Q * \mu * (L_1 - L_2)/[2 * \pi * (P_2 - P_1)] \tag{11.5.13}$$

where:

$L_1 = \ln (T + \Delta t_1) / \Delta t_1)$
$L_2 = \ln (T + \Delta t_2) / \Delta t_2)$.

The extrapolation of the pressure on a Horner plot yields the initial reservoir pressure (Figure 11.5.1).

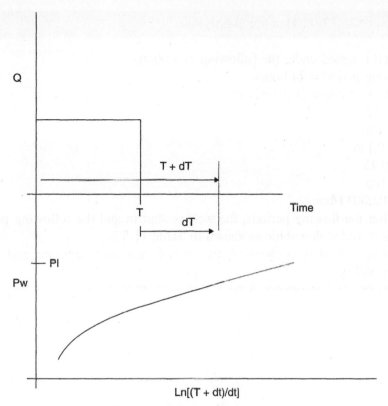

Figure 11.5.1 Pressure Buildup Analysis

Table 11.5.2
Example of pressure buildup

Time (hr)	Pressure (bar)
0	205.32
0.5	217.89
1.0	221.95
1.5	223.05
2	223.53
2.5	223.81
3	223.96
4	224.20
6	224.48
8	224.68
10	224.82
12	224.96

Exercise 11.4. Horner Analysis

A well is tested under the following conditions:

Flowing period = 24 hours

Production rate = 200 stm^3/day

B_o = 1.2

h = 10 m

r_w = 0.1 m

ϕ = 0.25

μ = 1 cp

C = 0.0003 1/bar.

After the flowing periods, the well is shut in and the following pressures recorded downhole as shown in Table 11.5.2.

Using the data, determine the initial reservoir pressure and the permeability.

HOMING-IN TECHNIQUES

Homing-in techniques are used when the position of one borehole relative to another needs to be known. Reasons why relative positions may be important are:

- In relief well drilling for blowouts, it may be necessary to intersect one borehole with another to enable the blowout well to be killed.
- Where survey data are unreliable, there may be a need to avoid collision between boreholes or to pinpoint the location of one well in relation to another.
- Some production/injection schemes require wells to be a fixed distance apart.

The principal methodology available for homing in comprises electromagnetic and magnetostatic techniques. While there has been research in the past on acoustic homing-in techniques, these have not been found to be successful.

12.1 MAGNETOSTATIC HOMING IN

Magnetostatic homing-in techniques use the fact that steel placed in a borehole usually has some remnant magnetization that causes a disturbance to the local magnetic field of the Earth. By running a sensitive magnetometer in open hole in a well close to another that has steel in it, this magnetic disturbance may be detected. Interpretation of the magnetic field detected as a function of depth can, in some cases, yield an accurate estimate of the distance and direction of the target well. Basic limitations on the usefulness of such methods are the generally short range over which such disturbances may be detected (typically <15 m) and the uncertain nature of the magnetization.

Since standard surveying tools, such as MWD (measurement while drilling), now possess accurate magnetometers, there is no need for dedicated tools for magnetostatic homing in. However, interpretation from such tools is not straightforward. It is hoped that the following will provide some useful techniques, as well as a better understanding of magnetic surveying in general.

12.1.1 Magnetization of Steel Casing/Drillstrings

As a result of the manufacturing process, or from subsequent magnetic inspection or mechanical shocks, the steel used in making casing, drillpipe, or accessories (e.g., jars, drill collars, bits) usually acquires a degree of magnetization. The simplest model for describing the field due to such a magnetization is in terms of a superposition of magnetic point monopoles of either north or south polarity. A north monopole is defined as one for which the lines of magnetic flux flow symmetrically and spherically toward the pole; for a south pole, the lines of flux flow similarly away from the pole. Although free monopoles are never observed in nature, in the vicinity of a pole, of which the opposite pole is far removed, the field will be dominated by that of the near pole, and the far pole may be ignored. When two or more poles are a similar distance from the point of measurement, the field is a linear superposition of the monopole field due to each of the individual poles. The magnetic field due to a monopole in a Cartesian reference system is as follows:

$$F_x = \mu r * M * x / (4 * \pi * r^3) \qquad\qquad (12.1.1)$$

$$F_y = \mu r * M * y / (4 * \pi * r^3) \qquad\qquad (12.1.2)$$

$$F_z = \mu r * M * z / (4 * \pi * r^3) \qquad\qquad (12.1.3)$$

where:
$r = \sqrt{(x_2 + y_2 + z_2)}$, in m
M = the pole strength, in Webers (Wb)
F_x, F_y, F_z are the field strengths, in Tesla (T)
μr is the relative magnetic permeability of the medium
x, y, z are the distances, in m, from the monopole to the measuring point along the x, y, and z axes.

It is usually assumed that μr is 1.0, so this term is commonly disregarded. The system used to define the x, y, and z axes is such that x is grid north, y is grid east, and z points vertically downward.

12.1.2 Interpretation of Magnetic Anomalies

When a three-axis magnetometer is run to measure the magnetic field, the field components B_x, B_y, and B_z in the Cartesian system are projected onto the three axes of the tool. In the tool's frame of reference, the z axis lies along the borehole and the x and y axes, as a result of tool rotation, are arbitrarily oriented in the plane (called the **sensor plane**) perpendicular to the borehole. If accelerometers are also present in the tool, the angle between the x axis of the tool (called the **toolface**) and the highside (HS) of the hole is also known, thus making it possible to convert the x and y components in the tool's reference system to that of a system whereby x lies in the HS direction, and the y magnetometer is 90 degrees clockwise from this in the sensor plane (called the highside right, or HSR, direction). Once this rotational transformation has been made, the readings of the magnetometer are said to be in the highside reference system. B_x and B_y are replaced by B_{hs} and B_{hsr}, respectively. B_z remains unchanged, although to avoid confusion will be termed B_{ax}. If the inclination and azimuth of the hole are known, a further transformation can be made to convert B_{hs}, B_{hsr}, and B_{ax} to B_x, B_y, and B_z in the Cartesian reference system. These quantities are illustrated in Figure 12.1.1.

The field measured by a magnetometer in the survey well is the vector sum of the fields due to any anomalies and to the Earth. In interpreting

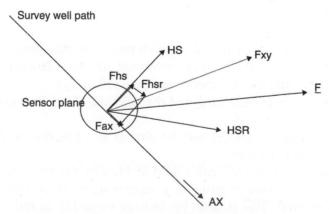

Figure 12.1.1 Fields in the Highside Reference System

any magnetic anomalies, the Earth's field must first be subtracted from the magnetometer readings. The basic procedure for locating the position of a target well, using a three-axis magnetometer combined with accelerometers, is therefore as follows:

1. Using the accelerometer data, the raw magnetometer readings are converted to the highside reference system.
2. The Earth's magnetic field is derived in the highside reference system $(E_{hs}, E_{hsr}, E_{ax})$ as a function of depth in the survey well. Note that while the Earth's field may be constant, the components in the survey well will vary if the well's inclination and azimuth vary.
3. The Earth's components are subtracted from the magnetometer components to derive the components of the disturbance (denoted as F). Hence:

$$F_{hs} = B_{hs} - E_{hs} \qquad (12.1.4)$$

$$F_{hsr} = B_{hsr} - E_{hsr} \qquad (12.1.5)$$

$$F_{ax} = B_{ax} - E_{ax}. \qquad (12.1.6)$$

Having calculated $\underline{F}$, it is convenient to derive the following quantities for comparison with theoretical models:

$$F_{xy} = \sqrt{\left(F_{hs}^2 + F_{hsr}^2\right)} \qquad (12.1.7)$$

$$F_{tot} = \sqrt{\left(F_{hs}^2 + F_{hsr}^2 + F_{ax}^2\right)} \qquad (12.1.8)$$

where
DF_{ax} = the rate of change of F_{ax} with respect to alonghole depth
HS_{dir} = the angle between $\underline{F}$, as projected into the sensor plane, called F_{xy}, and the HS (clockwise from HS to F_{xy})
AX_{dir} = the angle between $\underline{F}$ and the hole axis (from hole axis to $\underline{F}$).

The above five quantities may be plotted as a function of alonghole depth in the relief well.

4. The disturbance to the Earth's field, defined by the components of F, must be interpreted in terms of a superposition of monopoles along the target well. This is done by forward modeling as follows. From an assumed set of pole strengths and positions in terms of alonghole

depths in the target well, the magnetic field, in the highside reference system of the survey well magnetometer, is modeled. This involves using equations 12.1.1 to 12.1.3, together with survey data from both wells. In order to match the measured and modeled sets of data as functions of depth, it is necessary to move the assumed position of the target well with respect to the relief well and to vary the assumed distribution of poles on the target well until a best fit has been found.

Weaknesses in the above approach are as follows:

1. **Effect of smearing.** The point monopole model may not apply if the magnetization has become smeared along the axis of the pipe. The effect of this is that estimates of the distance made will be too large.
2. **Poorly defined HS.** If the relief well is nearly vertical, the HS direction ceases to be defined, and it is not possible to make the correct subtraction of the Earth's magnetic field in the sensor plane. This problem can be avoided by ensuring that the survey well is drilled with a few degrees deviation.
3. **Nonlinear well paths.** It is highly recommended that the survey well be drilled on a constant deviation and azimuth when passing close to the target. This is because normal surveying of the well path is not possible while the magnetometers are affected by the disturbance. While this problem could be solved through running a gyro in the survey well, this is not normally practical in open hole. Also, visualization of the survey and target wells is much harder if the highside reference system does not remain fairly constant.

12.1.3 The Earth's Field

It will now be shown how the components of the Earth's magnetic field in the highside reference system may be derived. Let the Earth's magnetic field be defined by:

- E_h = horizontal component of Earth's field
- E_v = vertical component of Earth's field
- Declination = angle clockwise from true north to magnetic north.

The Cartesian components of the Earth's field vector are given by:

$$E_x = E_h * \cos(\text{declination}) \tag{12.1.9}$$

$$E_y = E_h * \sin(\text{declination}) \tag{12.1.10}$$

$$E_z = E_v. \tag{12.1.11}$$

From the survey well deviation data it is possible, for each survey point, to define a Cartesian unit vector $\underline{\text{RVEC}}$, along the well trajectory. The HS direction is derived by first considering a unit vector 90 degrees clockwise from HS in the sensor plane, called $\underline{\text{HSR}}$.

$$\underline{\text{HSR}} = \begin{pmatrix} 0 \\ 0 \\ 1 \end{pmatrix} \wedge \underline{\text{RVEC}} \tag{12.1.12}$$

where $\wedge$ denotes the vector product (see Appendix 4). The HS vector becomes:

$$\underline{\text{HS}} = \underline{\text{HSR}} \wedge \underline{\text{RVEC}} \tag{12.1.13}$$

The three vectors $\underline{\text{HS}}$, $\underline{\text{HSR}}$, and $\underline{\text{RVEC}}$ must be converted to unit vectors (i.e., divided by their magnitude) to define the highside reference system. These are then denoted by $\underline{\text{HS}}^\wedge$, $\underline{\text{HSR}}^\wedge$, and $\underline{\text{RVEC}}^\wedge$. The Earth's components in the highside reference system are:

$$\underline{E}_{ax} = (\underline{E}.\underline{\text{RVEC}}^\wedge) * \underline{\text{RVEC}}^\wedge \tag{12.1.14}$$

where . denotes the scalar product.

$$\underline{E}_{hs} = (\underline{E}.\underline{\text{HS}}^\wedge) * \underline{\text{HS}}^\wedge \tag{12.1.15}$$

$$\underline{E}_{hsr} = (\underline{E}.\underline{\text{HSR}}^\wedge) * \underline{\text{HSR}}^\wedge. \tag{12.1.16}$$

12.1.4 Converting Survey Data to the Highside Reference System

The raw magnetometer tool components B_x, B_y, and B_z are converted to the highside reference system as follows. The survey tool's accelerometer readings will be A_x, A_y, A_z. The orientation of the toolface with respect to the HS direction is given by:

$$\theta = \arctan(A_y / A_x). \tag{12.1.17}$$

Table 12.1.1
Determining HSTF from accelerometer data

A_x	A_y	HSTF
+	+	$180 - \theta$
+	−	$180 - \theta$
−	+	$-\theta$
−	−	$-\theta$

The arctan function will normally return a value of θ between −90 and +90 degrees, and we are interested in the clockwise angle (between 0 and 360 degrees) between the HS and the toolface, denoted by HSTF. To derive HSTF from θ, the transformation must be applied as shown in Table 12.1.1.

B_{hs} and B_{hsr} are given by the following equations:

$$B_{hs} = B_x * \cos(\text{HSTF}) - B_y * \sin(\text{HSTF}) \qquad (12.1.18)$$

$$B_{hsr} = B_x * \sin(\text{HSTF}) + B_y * \cos(\text{HSTF}). \qquad (12.1.19)$$

The field due to a monopole, as measured in the highside reference system of a survey well, may likewise be modeled by replacing the components E_x, E_y, and E_z in the above equations with F_x, F_y, and F_z, as given by equations 12.1.1 to 12.1.3. In terms of the measurements made of F_{tot}, F_{xy}, F_z, ΔF_z, and AX_{dir}, the behavior will be as shown in Figures 12.1.2 to 12.1.4.

Similarly for a dipole, consisting of a north and a south pole of equal strength, the behavior will be as shown in Figures 12.1.5 to 12.1.7.

Note that in this example the axis of the dipole is parallel with that of the survey hole. If the survey hole passes closer to one pole than to the other, or if the poles are of different strengths, the behavior of the files will not be symmetric. In general, where more than one pole is present, it is necessary to model the field for different configurations and try to match with the measured data. This may be done by an automated procedure with a computer program.

12.1.5 Quicklook Interpretation Methods

Where the field is dominated by one pole, quicklook methods may be applied to estimate the shortest distance and direction to the pole.

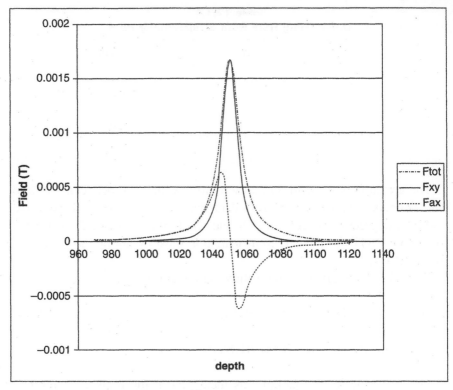

Figure 12.1.2 Field Due to a Monopole: F_{tot}, F_{xy}, F_{ax}

Consider a general case in which a survey well passes a target well with the closest approach distance of d (see Figure 12.1.8). Let a monopole be present on the target at a height of t above the point of closest approach.

The point at which the sensor is closest to the pole AX_{dir} will become 90 degrees. Note that this is not necessarily the point at which the wells themselves are closest together. Also, although one well passes the other, HS_{dir} remains constant with alonghole depth in the survey well. The shortest distance between the survey well and pole (x) may be determined from measuring the width of the F_{tot} curve at half the maximum intensity and dividing by 2. Hence:

$$x = \Delta F_{tot}/2. \tag{12.1.20}$$

Alternatively, the F_{xy} curve may be used, but the formula becomes:

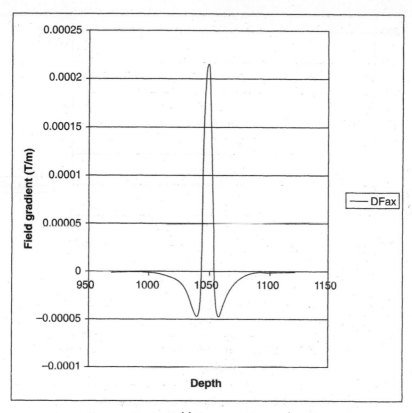

Figure 12.1.3 Field Due to a Monopole: ΔF_{ax}

$$x = \Delta F_{xy} * (0.652).\tag{12.1.21}$$

Similarly, the separation of the two stationary points (zero gradient) on the F_{ax} curve is related to x by:

$$x = \Delta F_{ax}/\sqrt{2}.\tag{12.1.22}$$

This may also be measured by finding the separation of the two zero crossing points of the ΔF_{ax} curve. If the separation of the two points on the AX_{dir} curve on either side of the point representing $AX_{dir} = 90$ degrees at which $AX_{dir} = 135$ degrees and 45 degrees is measured, x may be found from:

$$x = \Delta AX_{dir}/2.\tag{12.1.23}$$

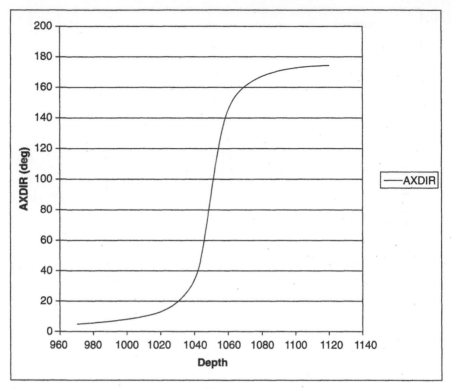

Figure 12.1.4 Field Due to a Monopole: AX_{dir}

In order to find the closest approach of the two wells, do as follows. Consider that the survey well has an inclination α. If the vector pointing toward the pole ($\underline{r}$) is an angle θ from the HS direction in the sensor plane, the angle between $\underline{r}$ and the horizontal is given by:

$$\phi = \arcsin([1 - \cos(\theta)] * \sin(\alpha). \tag{12.1.24}$$

The TVD (true vertical depth) of the pole is given by:

$$\mathrm{TVD}_{pole} = \mathrm{TVD}_{sensor} - x * \sin(\phi) \tag{12.1.25}$$

where the TVD of the sensor is at the point of closest approach to the pole.

The direction to the pole in the horizontal plane, φ_a, relative to the survey azimuth, is given by:

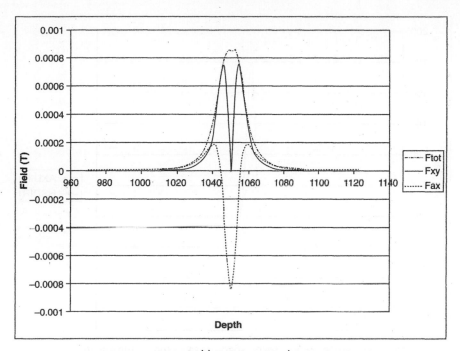

Figure 12.1.5 Field Due to a Dipole: F_{tot}, F_{xy}, F_{ax}

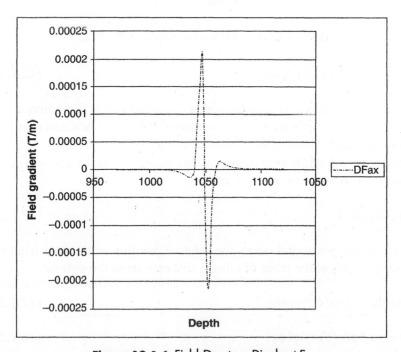

Figure 12.1.6 Field Due to a Dipole: ΔF_{ax}

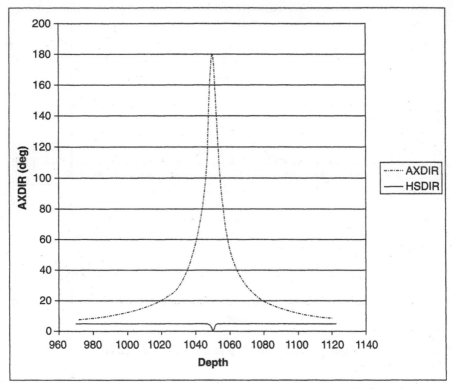

Figure 12.1.7 Field Due to a Dipole: AX_{dir}/HS_{dir}

$$\varphi_a = \arctan(\tan(\theta) * \cos(\alpha)). \tag{12.1.26}$$

Care should be taken to keep the sign of φ_a such that it is positive going clockwise from the horizontal projection of HS to the horizontal projection of the vector linking the sensor to the pole.

The direction of the pole relative to north, φ_n, if β is the azimuth of the survey well relative to north, is given by:

$$\varphi_n = \varphi_a + \beta. \tag{12.1.27}$$

Hence, in the horizontal plane, the relative position of the pole from the sensor position at the point of closest distance from the pole is:

north: $x * \cos(\phi) * \cos(\varphi_n)$ $\qquad$ (12.1.28)

east: $x * \cos(\phi) * \sin(\varphi_n).$ $\qquad$ (12.1.29)

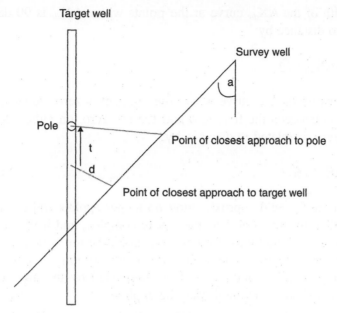

Figure 12.1.8 Survey Well Passing a Target Well

For dipoles, the situation is more complex, and a formula can be given for only the simplified situation of the survey and target wells being parallel and the separation of the poles being small compared with the separation of the wells. A dipole may be regarded as being positive if the north pole is above the south pole, and negative if the south pole is above the north pole. The polarity of the pole may be seen by observing AX_{dir} as the pole is passed. For a positive dipole, AX_{dir} increases with depth, reaching 180 degrees at the point of closest approach to the dipole, then decreasing again. For a negative dipole, AX_{dir} should decrease to zero as the dipole is passed, and then increase again. Since AX_{dir} is 0 or 180 degrees at the near point, HS_{dir} cannot be defined at this point. However, for a positive dipole, the direction to the target relative to HS is equal to HS_{dir} just above the near point. For a negative dipole, it is equal to HS_{dir} just below the near point. To find the Cartesian position of the dipole, equations 12.1.24 to 12.1.29 may be used.

Distance may be found from one of the following methods. The half-width of F_{tot} is related to the distance by:

$$x = F_{tot} * 0.453. \tag{12.1.30}$$

The width of the AX_{dir} curve at the points where AX_{dir} is 90 degrees is related to distance by:

$$x = \Delta AX_{dir}/\sqrt{2}. \qquad (12.1.31)$$

The points on the F_{ax} curve where the gradient is zero are such that the separation between the first zero and the last zero, on either side of the central zero, is related to distance by:

$$x = \Delta F_{ax}/\sqrt{6}. \qquad (12.1.32)$$

When the far field approximation no longer applies and the wells are not parallel, the situation becomes more complex, and it is not possible to use quicklook methods. However, it should be noted that at the point of closest distance from one of the monopoles forming a dipole, the field behavior of F_{tot}, F_{xy}, and F_{ax} will be dominated by the monopole, and monopole quicklook methods may be applied.

Exercise 12.1. Worked Field Example of Magnetostatic Homing In

Consider the following case. A vertical exploration well has blown out, and a relief well is required for homing in and intersection of the target well. The best survey data available for the target well are as follows:

Total depth: 2,200 ft TVDss

Easting: 5,340 ft

Northing: 6,898 ft

A relief well has been drilled to pass close to the blowout well, with the survey data in Table 12.1.2.

The Earth's magnetic field has the following components:

$$E_v = -7.75 * 10^{-6} \, T$$

$$E_h = 40.9 * 10^{-6} \, T.$$

Magnetic declination (the angle clockwise from grid north to magnetic north) is zero.

In the relief well, a survey package consisting of three magnetometers and three accelerometers has been run. The results are shown in Table 12.1.3 (note that depths/distances are in feet and not meters).

Table 12.1.2
Example of well survey data

TVD (z) (ft)	North (x) (ft)	East (y) (ft)
2073.4	6985.4	5348.8
2080.7	6975.9	5348.3
2086.7	6967.9	5347.9
2092.7	6959.9	5347.2
2098.7	6952	5346.8
2105.7	6941.6	5346.4
2111.7	6933.6	5345.9
2117.6	6925.6	5345.5
2123.7	6917.7	5345.1
2129.7	6909.7	5345.1
2135.8	6901.8	5344.7
2141.9	6893.9	5344.3
2148	6885.9	5343.8
2154.1	6878	5343.4
2161.2	6870.2	5343
2167.3	6862.4	5342.6
2173.5	6854.5	5342.1
2179.7	6846.7	5341.7

1. Derive the components of the Earth's field in the highside reference system.
2. Convert the raw magnetometer data to the highside reference system and subtract the Earth's field components.
3. Determine whether any poles are present, and their polarity.
4. Use quicklook methods to estimate the shortest distance from the pole to the relief well and estimate the error in the assumed position of the target well from the relief well.
5. Model the field due to the assumed pole in the highside reference system and overlay the field so derived with the actual measured data to see how good the fit is.
6. Is there any evidence of a second pole?

12.2 ELECTROMAGNETIC HOMING IN

Magnetostatic techniques have the advantage that dedicated tools are not required, but they rely on the unpredictable nature of the magnetiza-

Table 12.1.3
Example of raw magnetometer tool data

Depth (ft md)	A_x	A_y	A_z	B_x (μT)	B_y (μT)	B_z (μT)
2700	−1	0	1	−17.81	2.19	−34.76
2710	−1	0	1	−17.81	2.19	−34.76
2720	−1	0	1	−17.81	2.19	−34.76
2730	−1	0	1	−17.81	2.19	−35.37
2740	−1	0	1	−17.81	2.19	−35.37
2750	−1	0	1	−17.81	2.19	−35.37
2760	−1	0	1	−18.36	1.51	−36.59
2765	−1	0	1	−18.49	0.55	−37.8
2770	−1	0	1	−19.86	−1.51	−40.24
2772.5	−1	0	1	−21.23	−4.66	−41.46
2775	−1	0	1	−23.97	−8.77	−42.68
2777.5	−1	0	1	−26.71	−14.93	−40.85
2780	−1	0	1	−30.14	−19.73	−34.15
2782.5	−1	0	1	−30.14	−19.73	−28.05
2785	−1	0	1	−29.73	−12.88	−22.32
2787.5	−1	0	1	−28.77	−4.66	−23.17
2790	−1	0	1	−27.4	0.82	−25.61
2792.5	−1	0	1	−26.03	3.56	−29.88
2795	−1	0	1	−23.97	3.56	−31.71
2797.5	−1	0	1	−23.29	3.56	−33.29
2800	−1	0	1	−21.92	3.56	−34.15
2805	−1	0	1	−20.55	3.56	−34.51
2810	−1	0	1	−19.86	3.56	−34.51
2820	−1	0	1	−18.49	3.56	−34.15

tion occurring on the target. Also, the range is typically small. Electromagnetic techniques have a greater range, but they rely on dedicated tools, and the mathematical modeling is much harder to perform. The techniques have nonetheless been used successfully in a number of blowouts.

12.2.1 Principles of Electromagnetic Homing In

Continuous steel in a target well will provide a low-resistive path for any current induced into the formation. By measuring the intensity and direction of the magnetic field associated with the current on the target, the position of the target well may be determined. Hence, the main elements of an electromagnetic homing-in tool are as follows:

1. An electrode that injects a low-frequency current into the formation, separated by an insulated bridle (typically 100–300 ft long) from the sensor package
2. A sensor package consisting of orthogonal direct current (DC) and alternating current (AC) magnetometers. DC magnetometers are used to establish the tool orientation with respect to HS, and AC magnetometers are used to measure the magnetic fields caused by current on the target well.

The AC magnetic field measured in the sensor plane is the vector sum of the magnetic fields from each element of current. In the case of an infinite line source of current, the field measured at a sensor is given by Biot-Savart's law:

$$\underline{H} = I/[2*\pi*\underline{dist}]*[\underline{dist}^\wedge \wedge \underline{Tvec}^\wedge] \tag{12.2.1}$$

where:

$\underline{H}$ = vector of magnetic field strength, in T
I = current flowing, in amperes
$\underline{dist}$ = vector linking the sensor to the line source by shortest distance, in m
$\underline{dist}\wedge$ = unit vector along dist
$\underline{Tvec}\wedge$ = unit vector along the line source direction.

These components are illustrated in Figure 12.2.1.

If the current on the target well were everywhere constant, modeling would be relatively simple, and similar algorithms to those applied in the magnetostatic methodology could be applied. Unfortunately, this is not the case, since the current is influenced by the following factors:

1. Even in a theoretical case of isotropic media, and infinite target casing of uniform thickness, predicting the current as a function of depth in the target well is a complicated mathematical process involving the use of Bessel functions and numerical integration. The field measured in the survey well is affected not only by the sensor's proximity to the target, but by the injecting electrode's proximity to the target casing, which both vary with depth.
2. In reality the target casing is not infinite, and often one is homing in near the shoe of a casing string, where the target current will fall to zero (yielding no magnetic field).

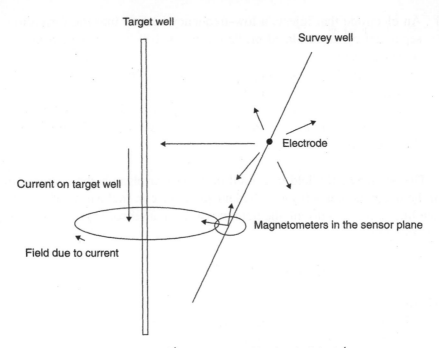

Figure 12.2.1 Electromagnetic Homing-in Principles

3. The current in the target is influenced by the thickness of the casing/drillpipe (which may vary with depth) and quality of the conducting path between steel and the formation, which is also variable.

In some field cases there have been attempts to overcome some of these limitations by injecting current directly into the target well at surface. While this has some advantages, it is nevertheless virtually impossible to accurately predict the current as a function of depth in the target well. During my time in research, I was involved in developing a series of Fortran programs that performed a full mathematical simulation of the tool response as a function of depth for deviated well paths, incorporating the effect of noninfinite target casing. While these programs are still available, it is beyond the scope of this book to explain the algorithms in detail.

In Figure 12.2.2, a typical response of the total field strength and HS_{dir} of the field as a survey well passes a target well are shown.

Note that the direction of the field (unlike that from a monopole) varies with depth as the survey well passes the target well. This means that while

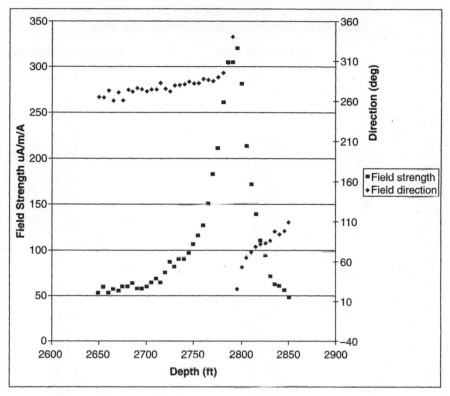

Figure 12.2.2 Typical Response of an Electromagnetic Homing-in Tool

modeling of the field strength may be very difficult, triangulation may also be used to help pinpoint the target well position.

12.2.2 Quicklook Interpretation Techniques

Consider a simple case of a straight vertical target well being passed by a straight deviated survey well having inclination α. The behavior of HS_{dir} as a function of depth will be given by:

$$HS_{dir} = \arccos\left[D * \cos(\alpha) \big/ \sqrt{(x^2 * \sin^2(\alpha) + D^2 * \cos 2(\alpha))} \right] \qquad (12.2.2)$$

where:
HS_{dir} = angle from highside to the measured field in the sensor plane
$\quad D$ = closest approach distance between survey well and target well
$\quad x$ = measured depth in the relief well, with $x = 0$ at closest approach point.

Table 12.2.1
Example of electromagnetic homing-in tool data

Depth (ft)	Field Direction (deg)	Field Strength (μa/m/a)
2650	265.46	52.96
2655	264.67	60.03
2660	273.54	53.32
2665	260.41	57.53
2670	270.72	55.05
2675	261.09	59.82
2680	274.32	59.73
2685	271.95	64.31
2690	275.94	58.05
2695	275.1	57.55
2700	272.5	60.77
2705	274.87	64.82
2710	274.72	69.11
2715	282	64.95
2720	275.21	75.95
2725	272.39	87.64
2730	279.19	82.18
2735	279.96	90.91
2740	280.95	91.24
2745	284.25	97.06
2750	282.18	107.72
2755	282.91	116.87
2760	287.81	127.97
2765	286.89	150.73
2770	285.39	183.33
2775	289.38	211.61
2780	295.55	262.2
2785	308.07	304.58
2790	341.45	305.65
2795	25.87	320.93
2800	52.92	281.65
2805	65.24	214.19
2810	72.56	172.37
2815	78.53	139.15
2820	82.29	111.05
2825	83.05	94.78
2830	86.01	71.93
2835	98.11	63.06
2840	94.1	61.25
2845	98.8	56.25
2850	109.64	48.43

At $x = 0$, HS_{dir} is also zero, which means that the field direction is along the HS direction. When $HS_{dir} = 45$ degrees and $x = X_{45}$, rearranging equation 12.2.2. yields:

$$X_{45} = D * \cot(\alpha). \tag{12.2.3}$$

Hence, if we plot HS_{dir} vs. x and measure the width Δ over which HS_{dir} varies from -45 to $+45$ degrees from its value at $x = 0$, we can say:

$$D = \tan(\alpha) * \Delta/2. \tag{12.2.4}$$

If the target well is not vertical, there will be a static shift in HS_{dir}, and α needs to be taken as the intersection angle between the survey well and target well. Intensity data may also be used for triangulation purposes if it can be assumed that the current on the target well is approximately constant as the sensor passes the target well. It is also necessary that any background signal be removed. This is done as follows. The separation in measured depth (Δ) between points for which the intensity curve has fallen to half its maximum height is given by:

$$\Delta = (D/\sin(\alpha)) * 2 * \sqrt{\left(y + \sqrt{(y^2 + 3)}\right)} \tag{12.2.5}$$

where $y = (2 - \cos^2(\alpha)) / \cos^2(\alpha)$ and D is the distance of closest approach.

Exercise 12.2. Interpretation of Electromagnetic Homing-in Data

A survey well was drilled at an inclination of 50 degrees past a vertical target well. After correction for background effects, an electromagnetic tool measured the data given in Table 12.2.1.

With the quicklook methods described above, measure the distance at the point of closest approach, using both the directional and intensity data.

WELL DEVIATION, SURVEYING, AND GEOSTEERING

13.1 WELL DEVIATION

The trajectory of a deviated well may be described in terms of its inclination, depth, and azimuth. The **inclination** of a well at a given depth is the angle (in degrees) between the local vertical and the tangent to the wellbore axis at that depth (Figure 13.1.1). The convention is that 0 degrees is vertical and 90 degrees is horizontal. Parts of a degree are given in decimals, rather than minutes and seconds. Gravity varies with latitude, and its direction may be influenced by local features such as mineral deposits and mountains, as well as the Earth's rotation.

Depth in boreholes is measured either along the hole itself, in which case it is referred to as measured or alonghole depths, with reference to a fixed point, or as true vertical depth (TVD) with reference to a datum. Depth references that are commonly used are as follows:

- *Derrick floor.* This is the elevated deck on which the rig crew work, typically 10 m or so above ground level on a land rig and 20–30 m on an offshore rig. Also sometimes referred to as a rotary table.
- *Kelly bushing.* This is the top of the bushing, which rotates on the derrick floor (although kellys are rarely used on modern drilling rigs with topdrives) and is typically 1 ft higher than the derrick floor.
- *Mean sea level.* This is the elevation of the sea, averaging out the effect of tides or seasonal variations. Usually the topography department will establish the elevation of a land location prior to drilling. For offshore locations, the elevation of the seabed will be known. On floating rigs, a correction using tide tables will be used.

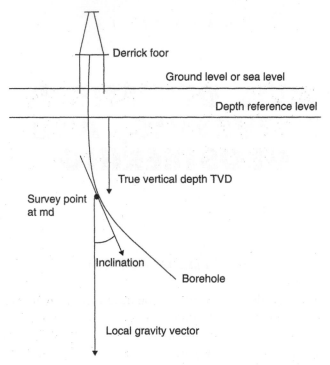

Figure 13.1.1 Basic Definitions

Azimuth, expressed in degrees between 0 and 360, is defined as the angle of the horizontal component of the direction of the wellbore at a particular point measured in a clockwise direction from magnetic north, grid north, or true north. These are defined:

- *Magnetic north.* This is the direction of the horizontal component of the Earth's magnetic field lines at a particular point on the Earth's surface. A magnetic compass will align itself to these lines. The angle between magnetic north (MN) and true north (TN) is defined as the magnetic declination (D). When MN lies to the west of TN, D is negative. When MN lies to the east of TN, D is positive. Typical values for D are −5 to +5 degrees.
- *Grid north.* Due to the curvature of the Earth, it is not possible to cover the surface in a regular rectangular grid pattern using meridians (i.e., lines heading north/south), although such a grid will be almost rectangular over limited areas. The central meridian in a grid will be identical to true north, but vertical grid lines to the west of center will point

west of true north in the Northern Hemisphere and east of true north in
the Southern Hemisphere. Likewise, vertical grid lines to the east of the
central meridian will point east of true north in the Northern Hemisphere and west of true north in the Southern Hemisphere. The so-called
grid correction (G) is positive when TN is east of grid north, and negative when it is west. Typical values are -1.5 to $+1.5$ degrees.
- *True north.* This is the direction of the geographic north pole as defined
  by the axis of rotation of the Earth. The meridians, or lines of longi-
  tude, on maps point toward true north.

13.2 SURVEYING

Borehole position uncertainty defines the range of actual possible positions of a particular point in a well in terms of eastings, northings, and
TVD. Factors that affect borehole position uncertainty are:

1. **Accuracy of measured depth determination in the well.** Both
   drillpipe and wireline cable suffer from stretch and inaccuracies in the
   methods used to measure how much pipe or cable has been run into
   the hole. This uncertainty becomes greater with increasing depth and
   well deviation. For a vertical well drilled to 3,500 m, one would expect
   the measured depth at total depth (TD) to be known to an accuracy of
   roughly 2 m. For a deviated well drilled to 3,500 m TVD with a deviation at bottom of 50 degrees, this inaccuracy might rise to 5 m.
2. **Frequency of survey stations.** When surveying a well, it is normal to
   acquire "stations" at discrete depth intervals along the well. At each of
   these stations, the hole's inclination and azimuth will be measured.
   Between stations, it is necessary to use an algorithm to determine
   values to be interpolated. The accuracy of the final survey will depend
   on the frequency of these stations and algorithm used.
3. **Survey tool accuracy.** Let us consider these different types of tools
   separately:
   (a) *Magnetic survey tools.* The accuracy of magnetic devices is limited
       by their intrinsic accuracy and the extent to which they are affected
       by magnetic interference. The strength and direction of the Earth's
       field is also a factor, since in a worst-case scenario of drilling in
       the same direction as the Earth's lines of flux, no azimuthal mea-
       surement would be possible. Near the poles, the Earth's field is
       nearly vertical and this would be a factor. Magnetic interference
       may come from any metal in proximity to the tool (such as the

drillstring itself, although nonmagnetic collars are always used immediately beside the tool). The tool's intrinsic accuracy is affected by the sensitivity of the magnetometers and accelerometers (used to determine tool orientation). From published data it is estimated that lateral borehole uncertainty arising from use of properly calibrated magnetic survey tools is on the order of 14 m per 1,000 m in a vertical well and 20 m per 1,000 m in a 70-degree deviated well.

(b) *Gyro survey tools.* These types of tools are affected by drift in the alignment of the gyro orientation during the survey. They are usually run only when casing has been set in a well, so they cannot be used for decision making during the course of drilling. From published data it is estimated that lateral borehole uncertainty arising from use of properly calibrated gyro survey tools is on the order of 1.5 m per 1,000 m in a vertical well and 8 m per 1,000 m in a 70-degree deviated well.

(c) *FINDS.* These tools use highly accurate accelerometers and double-integrate the accelerations to determine absolute distance moved by the tool during the survey. Their accuracy is estimated at 0.5 m per 1,000 m irrespective of deviation.

13.2.1 The Effects of Borehole Position Uncertainty

We need to know the positions of wells accurately for the following reasons:

1. **Well safety.** In the event of a well blowing out, a relief well may be required to intercept or pass close to the well. Homing in with a relief well (see Chapter 12) is made easier if the position of the target well is known accurately. Also, for well collision avoidance in densely drilled areas, the relative well positions must be known.

2. **Mapping.** Any geological maps are only as good as the input data. While small uncertainty in the lateral position of wells may not be critical, uncertainty in the TVD at which a certain horizon is penetrated may lead to serious errors in maps, and therefore reserves. In particular, if the depth of a fluid contact appears to be different in neighboring wells, it might lead one to make wrong judgments as to the position of faults or communication/differential depletion between wells.

3. **Geosteering.** When drilling horizontal wells through thin horizons, accurate measurement of TVD is of high importance.

4. **Pressure/gradient determination.** Accurate knowledge of the TVD at which pressure measurements are made is essential for accurate determination of gradients and correction of formation/well pressures to a common datum reference.
5. **Legal reasons.** When drilling close to a concession or national boundary, it may be essential to avoid accidentally crossing such a boundary. There may also be implications for equity determination and unitization. Most government bodies will have minimum requirements regarding survey accuracy and maintenance of a proper database of survey data.

13.3 GEOSTEERING

Geosteering is the use of information gained while drilling to make real-time decisions on the trajectory of the well. Such decisions may be essential to optimize the utility of a well. Geosteering is used (a) in high-angle deviated wells in thin formations where productivity can be achieved only if the wellbore remains in a thin permeable zone and (b) in horizontal wells where it is necessary to remain a fixed distance from either a fluid contact or an overlying tight formation, as well as during (c) drilling in close proximity to a fault where it is necessary to establish whether or not the fault is close and should be crossed and (d) drilling with a fixed orientation to natural fractures.

Data that may be used in the decision-making process during geosteering concern (1) deviation; (2) cuttings, including hydrocarbon shows and gas readings; (3) transmission of LWD (logging while drilling) tools in real time, typically up/down GR (gamma ray), density, neutron, and resistivity; and (4) drilling parameters, such as losses, kicks, rate of penetration (ROP), and torque.

Geosteering is often much harder in practice than anticipated, due to the following factors:

- Tools used in the decision-making process are typically run some way behind the bit (possibly up to 30 m). Therefore, if the bit is not where you want it to be, you will often not know about it until quite a bit of formation has been penetrated.
- In high-angle wells, there are often problems with real-time data transmission through mud pulses arising from noise, high ROP, tool failures, battery life limitations, and bandwidth.
- Cuttings data may take up to 2 hours to reach surface (the "bottoms-up" time). Where a turbine is used, the cuttings may be very finely

ground and difficult to interpret. Also, highly deviated wells are often drilled with OBM (oil-based mud), making hydrocarbon differentiation difficult.

- Areal variation in the formation is usually much greater than that expected from the working geological maps. It is very often the case that subseismic faults of a few meters are encountered, which cause the well to suddenly go out of the target zone. Often it is not clear whether one has exited the top or base of the target zone, so one does not know whether to drill up or down to get back in. Even where faulting is not present, there may be thinning or deterioration of reservoir properties that were not envisaged.

- Even where the right geosteering decisions can be made, control of deviation in the well itself may be a problem. When one is entering a thin horizon at a steep angle, it may be impossible to avoid immediately exiting the horizon on the other side. There may also be a tendency for the bit to drop or turn to the right or left, which cannot easily be controlled. In very long horizontal wells, one may be limited by the need to keep the drillpipe in tension and have sufficient weight on bit (WOB) to be able to make further progress.

- Where a horizontal well accidentally penetrates a water-bearing zone, there may be significant practical difficulties in preventing a large proportion of the well's production from originating in the water zone. The possibilities of isolating certain zones in long horizontal wells are very limited.

In spite of the above limitations, geosteering can be immensely valuable in drilling very highly productive wells and can make the difference between a field being economically viable or not. It may also be the case, if drilling in a permeable formation surrounded by tight formations or in a long horizontal well, that the bit will naturally follow a path of least resistance and steer itself within the most permeable layer, effectively "bouncing off" the harder layers. An example of a typical geosteered well through a thin formation is shown in Figure 13.3.1.

With respect to the decisions made by the petrophysicist in the planning and execution of a geosteered well, it is worthy of consideration that while one would ideally want as many tools in the hole as possible, with both up and down measurements of all parameters, one is necessarily limited by constraints as to what the drillers are prepared to have in the toolstring (a greater number of tools and their proximity to the bit affect drillers' ability to steer the well) and what data can be captured within the

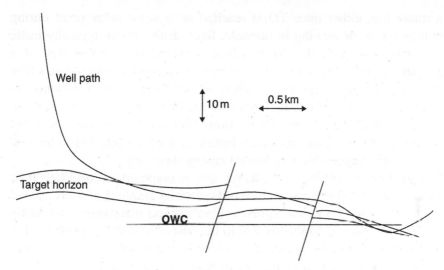

Figure 13.3.1 Example of Geosteered Well

available bandwidth (of the mud pulse telemetry system). Therefore, careful consideration should be given to which tools are most effective in determining whether or not one is in the target formation as opposed to above or below it.

Bear in mind that the density/neutron tools require the toolstring to be rotating for meaningful data to be obtained, and when changing the well course it is often necessary to slide the toolstring using a turbine and bent-sub. Resistivity data are generally more reliable, since they are not a statistical type of measurement. The LWD-GR devices can generally be placed closer to the bit and may be sufficient in many cases for determining whether one is exiting a target formation from above or below.

If a long bit run is planned, battery life may be an issue (typical battery life is 50–100 hours), as may the downhole memory in which data, assumedly, are being recorded, which may become full after a certain number of hours. It is generally recommended to always record the data in a downhole memory in addition to pulsing to surface. To avoid making additional runs with pipe-conveyed logging at TD, it may be considered worthwhile to include tools in the toolstring set to only record downhole and not pulse to surface.

When permeability or presence of fractures is a particular issue, there may be a requirement for tools (such as NMR [nuclear magnetic resonance], pressure testing, or sonic) that are not available from all the contractors. Data that are missing or of poor quality may be reacquired during a round trip, either once TD is reached or at some other point during pulling out of or running in the hole. Such decisions are typically made in conjunction with the drillers. It is recommended at the start of a geosteered well to set up a strict and rigorous system of naming data files transmitted from the rig to the office, so that there is no confusion as to whether data are pulsed or memory, and whether or not they were acquired during drilling or tripping. Distinctions also need to be made between up/down data and data for which depths have been corrected to be consistent with previous runs or known casing shoes, etc.

Tool failures during geosteering are a common occurrence. It is recommended to keep accurate records of serial numbers of tools used in the hole and to check that regular calibration and maintenance are being performed. It can significantly add to the cost of a well if a toolstring has to be brought back to surface due to tool failure in the critical section of a well, and in some cases may even result in the well being lost if the openhole time becomes too great.

For a geosteered well to be successful, there needs to be good communication between the petrophysicist, wellsite geologist, office geologist, and drilling department. The wellsite geologist, particularly if he has a good knowledge of the field, is usually in the best position to know which formation the well is in, but he needs the support of the petrophysicist to interpret the real-time formation evaluation data. The necessary course of action that these two decide upon needs to be fed back to the drillers so that the well trajectory is optimized.

The depth offset between the up and down readings of tools can be used to establish whether the trajectory is veering structurally deeper or shallower. This is done as follows. Consider that one is drilling a low-GR sand, bounded above and below by high-GR shales. In the event that one exits the sand into the structurally shallower shale, one would expect that the up reading on the GR would respond before the down reading; similarly, if the wellbore exits to the structurally deeper shale, the down log would respond first. The offset between up and down readings, together with knowledge of the borehole size, can yield an estimate of the relative dip between the borehole and formation. Consider the example in Figure 13.3.2.

If the borehole diameter is d and the offset between the up and down readings is t (measured in similar depth units as d), then the relative angle (θ) between the borehole and the formation is given by:

$$\theta = \arctan(d/t). \tag{13.3.1}$$

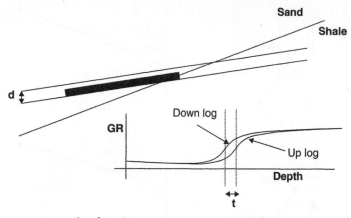

Figure 13.3.2 Example of Up/Down Response as Borehole Crosses Boundary from Above

If the borehole deviation is ϕ, the apparent formation dip (α) will be given by different formulae depending on whether the wellbore leaves by the top or bottom of the formation, and whether or not the formation dip is broadly in the same direction or opposite to that of the borehole. The appropriate formulae are:

Case A. Exit by bottom, formation dip opposite to borehole:

$$\alpha = (\phi - 90) + \theta \qquad\qquad (13.3.2)$$

Case B. Exit by top, formation dip opposite to borehole:

$$\alpha = (90 - \phi) + \theta \qquad\qquad (13.3.3)$$

Case C. Exit by bottom, formation dip same direction as borehole:

$$\alpha = (90 - \phi) - \theta \qquad\qquad (13.3.4)$$

Case D. Exit by top, formation dip same direction as borehole:

$$\alpha = (90 - \phi) + \theta \qquad\qquad (13.3.5)$$

These four scenarios are illustrated in Figure 13.3.3.

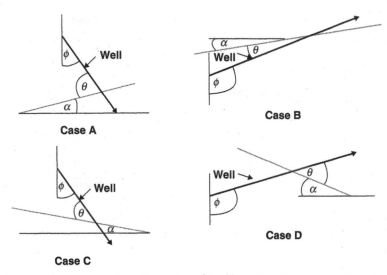

Figure 13.3.3 Four Scenarios of Wellbore Leaving a Formation

If there is an angle between the borehole azimuth and formation dip direction, the true formation dip will be greater than α. In order to correct for this, the following formula should be used:

$$\tan(\Delta) = \tan(\alpha)/\cos(\gamma) \qquad\qquad (13.3.6)$$

where:
Δ = the true formation dip
α = the apparent formation dip
γ = the angle between the azimuth of the wellbore and the maximum dip direction.

Note that in the event that one is drilling along the strike of the formation (i.e., $\gamma = 90$ degrees), it is not possible to say what the true formation dip is.

Exercise 13.1 Formation Dip from Up/Down Logs

One is drilling an $8^{1}/_{2}$-in.-diameter hole at a deviation of 95 degrees when the reservoir is exited. The offset between the up and down readings is 2 m, with the up reading responding first.

1. What is the relative dip between the borehole and formation?
2. If the direction of dip of the formation is the same as the borehole, what is the absolute formation dip?
3. Suppose that it is known that the formation dip azimuth is at an angle of 40 degrees to the borehole trajectory. What is now the true formation dip?

13.4 HORIZONTAL WELLS DRILLED ABOVE A CONTACT

Often there is a requirement to drill wells a fixed distance above a water contact in order to optimize drainage. In an ideal reservoir, which is homogeneous, the contact would be at a fixed subsea depth, so in theory one would only need to keep the well at a certain TVD. In practice, contact depths may vary over a reservoir due to:

- *Capillary effects.* If the rock quality (particularly permeability) varies, the oil/water contact (OWC) or gas/water contact (GWC) will vary, while the free water level (FWL) remains constant.
- *Depletion in the field.* The contact may have moved due to aquifer influx or injection during production.
- *Depletion in neighboring fields.* There may be observed an overall tilting of the contact in a certain direction due to offtake in a neighboring field affecting the aquifer.

The position of the contact will typically have been determined through measurements made in nearby wells, and there may be some scatter in the interpreted contact depths due to surveying errors. Typically, this uncertainty will be on the order of 2–5 m, although it may be greater if some wells are particularly anomalous. Borehole TVD uncertainty as a result of surveying errors, coupled with uncertainty in the true contact depth, will lead to an overall uncertainty as to the distance between the well and the contact. It may therefore be necessary for the petrophysicist to assess the well's proximity to the contact or FWL via real-time measurements during drilling. The best way to do this is by using an established saturation/height function, calibrated against core in earlier wells. Then the water saturation calculated in the horizontal well while drilling may be input to the model to back-calculate the height above the FWL. Once this is known, the OWC may be estimated by observing the entry height on the curve corresponding to the prevailing porosity and permeability.

Using the equations presented in Chapter 4 and solving for h (the height above FWL) yields:

$$h = \left\{ \left[(S_w - S_{wirr})/a \right]^{(-1/b)} \Big/ \sqrt{(k/\phi)} \right\} \Big/ \left[(\text{rho}_w - \text{rho}_h) * 3.281 * 0.433 \right] \quad (13.4.1)$$

where:
h = the height above FWL, in m
S_w = log-derived saturation (fraction)
S_{wirr} = irreducible water saturation (fraction)
a, b = Leverett J fitting constants
k = permeability, in md
ϕ = porosity (fraction)
rho$_w$ = water density, in g/cc
rho$_h$ = hydrocarbon density, in g/cc.

It would be necessary to derive an empirical relationship for the height between the OWC (or GWC) and FWL and to have a log-derivable parameter (such as permeability from the poroperm relationship). In Figure 13.4.1 is a hypothetical example of a well drilled close to an FWL, using the log-derived saturation to estimate the depth of the FWL.

In the figure, production from neighboring fields was known to cause a variation in the FWL. Entry height between the FWL and OWC was

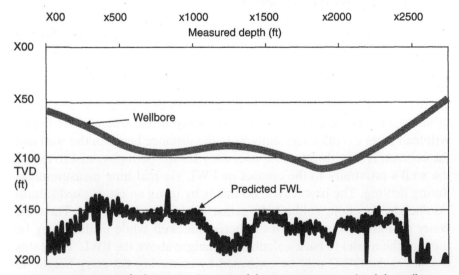

Figure 13.4.1 Calculating the Position of the Free Water Level While Drilling

known to be small. Note that the small-scale variation in the FWL depth is probably not real but due simply to inaccuracy/scatter in the calculated S_w. Also the fact that the FWL appears to go much deeper as the wellbore rises is almost certainly spurious and a result of inaccuracy in the saturation/height model. However, such a plot could at least be used to confirm a general dipping of the FWL along the wellbore trajectory and to establish that the well was some 50 ft above the contact.

13.5 ESTIMATING THE PRODUCTIVITY INDEX FOR LONG HORIZONTAL WELLS

A technique that has often been found to give good results in estimating PI is as follows. Using the established poroperm relationship, determine k all along the horizontal wellbore. Integrating this function from the top of the objective to TD will yield the gross product $k*h$ for the well. Make a graph of $k*h$ vs. the PI (in b/d/psi drawdown) for a number of wells already producing in the area. Often a good correlation is found. This enables one to predict the PI while the well is being drilled, taking into account sections of the well that will be nonproducing due to poor permeability. This information may be important because the total length a well needs to be drilled may be shortened (thereby saving money) if a PI threshold has been reached above which surface facility limitations will negatively affect production. It allows comparisons to be made between wells drilled under different conditions and helps identify formation damage in wells that produce far below the established trend.

TEST WELL 1 DATA SHEET

DEPTH	GR	DENSITY	NEUTRON	RES_DEEP	RES_SHAL	RES_MICR	CAL	DT	DTS
616.001	104.638	2.663	0.129	19.841	21.747	13.946	9.277	71.991	162.778
616.153	102.528	2.654	0.131	20.287	22.23	16.053	9.259	71.613	161.923
616.306	99.254	2.634	0.128	20.183	22.155	19.915	9.241	70.83	160.152
616.458	97.172	2.648	0.124	19.741	21.774	22.498	9.205	70.163	158.646
616.61	95.056	2.664	0.118	19.241	21.26	21.176	9.169	69.291	156.674
616.763	90.259	2.639	0.11	18.752	20.65	19.21	9.151	68.946	155.893
616.915	88.342	2.613	0.107	18.612	20.471	18.568	9.097	69.323	155.299
617.068	88.537	2.638	0.107	18.705	20.519	20.157	9.062	69.294	155.402
617.22	89.109	2.664	0.11	18.581	20.272	20.543	9.008	69.41	156.159
617.372	87.462	2.683	0.112	18.396	20.045	19.703	8.936	69.671	155.317
617.525	86.037	2.671	0.111	18.327	20.01	19.32	8.918	73.536	162.649
617.677	89.761	2.641	0.113	18.215	19.898	20.959	8.918	76.975	173.815
617.83	94.643	2.652	0.112	16.986	18.492	20.651	8.918	74.788	169.102
617.982	97.146	2.651	0.109	17.263	18.692	18.447	8.936	71.832	162.418
618.134	94.166	2.637	0.105	19.81	21.735	16.25	8.936	64.523	145.893
618.287	89.936	2.629	0.104	20.603	23.073	16.234	8.936	58.767	132.831
618.439	87.886	2.636	0.105	20.175	22.72	17.906	8.954	59.427	132.792
618.592	89.367	2.643	0.11	20.086	22.624	16.82	8.972	62.615	141.077
618.744	90.908	2.639	0.117	19.978	22.564	16.566	8.972	66.49	150.34
618.896	90.69	2.653	0.125	19.667	22.313	19.505	8.918	68.539	154.973
619.049	88.706	2.666	0.125	19.177	21.781	21.885	8.882	68.012	152.672
619.201	87.438	2.677	0.117	18.944	21.532	21.834	8.882	66.741	148.766
619.354	88.737	2.676	0.112	18.952	21.597	20.027	8.864	66.741	149.845
619.506	89.631	2.659	0.108	18.971	21.569	17.955	8.828	67.38	152.037
619.658	87.758	2.636	0.105	19.672	22.196	16.724	8.811	67.902	151.621
619.811	80.711	2.632	0.103	21.349	24.062	15.974	8.811	69.199	148.701
619.963	74.763	2.616	0.105	23.062	26.109	15.909	8.828	70.222	146.252
620.116	71.779	2.616	0.106	24.493	27.926	18.621	8.882	70.338	144.264
620.268	72.245	2.65	0.101	25.87	29.99	21.6	8.972	70.859	145.679
620.42	75.435	2.649	0.094	26.751	32.015	22.549	9.115	70.975	148.337
620.573	75.969	2.657	0.091	25.661	31.747	25.408	9.097	70.831	148.447
620.725	74.33	2.65	0.101	24.482	30.546	20.963	8.972	70.042	145.551
620.878	74.164	2.636	0.114	25.18	31.35	13.826	8.972	68.771	142.787
621.03	76.349	2.627	0.125	26.063	32.496	11.819	8.972	68.134	143.077
621.182	80.955	2.64	0.131	26.335	32.811	11.81	8.972	67.232	144.663
621.335	85.094	2.714	0.136	26.703	33.219	11.769	9.008	65.81	144.809
621.487	87.045	2.742	0.147	27.91	34.409	11.782	8.972	63.387	140.982

DEPTH	GR	DENSITY	NEUTRON	RES_DEEP	RES_SHAL	RES_MICR	CAL	DT	DTS
621.64	87.427	2.712	0.156	27.974	34.027	15.187	8.864	62.762	139.888
621.792	86.204	2.69	0.159	24.524	29.497	19.316	8.811	63.901	141.466
621.944	87.928	2.665	0.156	21.336	25.55	24.765	8.793	64.91	145.077
622.097	92.263	2.637	0.147	20.516	24.586	33.757	8.775	63.992	144.693
622.249	93.768	2.62	0.131	20.467	24.482	49.697	8.739	62.508	141.337
622.402	83.772	2.598	0.105	20.321	24.25	69.502	8.667	63.405	138.514
622.554	65.717	2.574	0.078	19.522	23.279	68.768	8.631	64.162	127.65
622.706	51.685	2.562	0.06	17.837	21.234	67.897	8.685	64.133	119.312
622.859	48.366	2.559	0.054	15.861	18.899	66.04	8.757	64.795	118.721
623.011	49.786	2.544	0.054	14.257	17.044	40.424	8.775	64.294	118.57
623.164	48.655	2.543	0.053	13.042	15.576	23.724	8.793	63.147	115.854
623.316	44.275	2.527	0.051	12.543	14.94	15.228	8.738	63.523	114.263
623.468	40.645	2.483	0.051	12.638	15.064	11.226	8.631	63.872	113.054
623.621	39.411	2.483	0.052	12.73	15.163	12.217	8.595	63.93	112.546
623.773	39.574	2.483	0.055	12.529	14.888	11.607	8.613	63.872	112.524
623.926	41.373	2.474	0.058	12.116	14.387	10.873	8.631	68.759	122.096
624.078	42.306	2.485	0.063	11.852	14.032	10.875	8.595	73.263	130.631
624.23	42.094	2.491	0.064	11.34	13.327	10.854	8.559	73.148	130.304
624.383	40.769	2.485	0.064	10.306	12.021	9.294	8.577	72.59	128.555
624.535	38.563	2.485	0.062	9.68	11.219	8.593	8.613	64.172	112.554
624.688	37.966	2.493	0.06	9.744	11.122	9.476	8.667	58.673	102.641
624.84	37.696	2.482	0.059	10.796	12.115	8.02	8.667	59.812	104.512
624.992	37.4	2.457	0.058	14.675	16.298	8.029	8.649	57.983	101.186
625.145	36.744	2.464	0.058	18.261	20.048	9.661	8.613	55.43	96.456
625.297	34.4	2.492	0.057	17.63	19.091	9.631	8.541	55.943	96.372
625.45	31.687	2.547	0.055	16.332	17.641	9.652	8.541	62.623	106.641
625.602	30.038	2.637	0.056	15.087	16.381	9.621	8.559	68.132	115.218
625.754	30.352	2.695	0.058	13.219	14.393	9.593	8.577	69.895	118.354
625.907	31.492	2.724	0.06	11.14	12.118	9.617	8.595	73.8	125.57
626.059	31.599	2.725	0.06	10.995	11.796	10.435	8.595	70.092	119.315
626.212	30.805	2.651	0.062	13.139	13.839	14.329	8.595	70.308	119.281
626.364	29.543	2.532	0.066	14.572	15.224	11.157	8.595	76.598	129.264
626.516	28.825	2.478	0.072	15.176	15.718	7.811	8.559	76.069	127.986
626.669	30.027	2.471	0.075	16.415	16.735	8.591	8.524	75.611	127.859
626.821	32.016	2.467	0.072	16.839	16.982	8.04	8.541	71.205	121.424
626.974	33.514	2.484	0.068	16.583	16.573	7.794	8.559	64.393	110.509
627.126	33.251	2.472	0.063	16.627	16.397	7.77	8.559	63.029	108.046
627.278	32.073	2.461	0.062	16.73	16.28	7.76	8.559	63.291	107.954
627.431	32.169	2.472	0.061	16.561	15.947	7.754	8.559	62.507	106.66
627.583	33.04	2.471	0.063	16.814	16.067	7.67	8.559	62.248	106.611
627.736	34.388	2.495	0.065	17.38	16.511	9.406	8.559	62.364	107.427
627.888	34.497	2.513	0.065	18.509	17.487	9.97	8.559	61.987	106.827
628.04	33.604	2.502	0.064	19.368	18.198	9.782	8.577	62.246	106.865
628.193	33.841	2.485	0.061	18.653	17.498	10.725	8.595	63.146	108.52
628.345	34.574	2.477	0.06	17.214	16.192	10.573	8.595	63.669	109.762
628.498	34.954	2.492	0.062	15.953	15.132	10.927	8.595	63.93	110.392
628.65	34.182	2.495	0.065	15.136	14.468	10.94	8.595	64.681	111.319
628.802	33.892	2.49	0.066	14.577	13.972	9.945	8.613	66.329	114.015
628.955	34.468	2.488	0.065	14.288	13.705	8.663	8.613	67.235	115.858
629.107	33.92	2.482	0.065	14.2	13.614	6.92	8.595	65.949	113.376
629.26	32.146	2.461	0.066	13.612	13.142	4.277	8.577	64.041	109.268
629.412	31.748	2.418	0.07	9.292	9.208	2.943	8.577	63.668	108.447
629.564	32.748	2.373	0.077	5.606	5.772	2.726	8.595	64.422	110.198
629.717	33.523	2.379	0.086	5.114	5.438	2.719	8.595	64.538	110.762
629.869	33.001	2.399	0.095	5.804	6.267	2.717	8.595	64.423	110.318

DEPTH	GR	DENSITY	NEUTRON	RES_DEEP	RES_SHAL	RES_MICR	CAL	DT	DTS
630.022	32.212	2.391	0.1	6.157	6.634	2.674	8.613	64.423	109.949
630.174	32.689	2.393	0.097	6.18	6.648	2.382	8.631	64.045	109.527
630.326	33.32	2.403	0.09	6.165	6.66	2.264	8.631	64.563	110.709
630.479	33.59	2.407	0.083	6.165	6.666	3.371	8.631	65.582	112.586
630.631	33.86	2.415	0.077	6.154	6.641	8.461	8.631	65.582	112.716
630.784	32.415	2.487	0.075	6.02	6.495	12.602	8.595	65.205	111.381
630.936	30.504	2.513	0.078	5.685	6.151	10.32	8.649	65.205	110.484
631.088	29.894	2.457	0.087	5.203	5.669	10.025	8.757	65.322	110.397
631.241	30.772	2.42	0.096	4.902	5.301	9.995	8.775	65.582	111.248
631.393	32.806	2.411	0.099	5.393	5.71	9.966	8.775	66.104	113.103
631.546	33.12	2.421	0.1	6.41	6.769	9.951	8.775	67.258	115.232
631.698	33.096	2.409	0.101	7.214	7.665	6.476	8.775	68.018	116.522
631.85	34.478	2.408	0.102	7.892	8.378	3.365	8.757	67.612	116.512
632.003	35.788	2.438	0.102	8.637	9.18	3.496	8.703	66.71	115.607
632.155	34.341	2.454	0.098	9.588	10.172	4.546	8.631	66.076	113.798
632.308	31.345	2.471	0.088	10.76	11.272	4.487	8.631	65.814	111.912
632.46	29.325	2.485	0.08	11.255	11.704	4.432	8.649	65.437	110.329
632.612	30.689	2.479	0.078	11.681	12.146	4.436	8.631	65.841	111.649
632.765	34.057	2.459	0.079	13.261	13.768	4.43	8.631	66.104	113.709
632.917	36.446	2.434	0.082	15.034	15.556	4.427	8.631	65.844	114.43
633.07	39.231	2.442	0.083	19.206	19.609	4.434	8.631	65.96	116.027
633.222	40.526	2.462	0.084	29.714	30.148	4.435	8.631	65.96	116.688
633.374	39.886	2.462	0.083	35.857	37.272	5.129	8.631	65.96	116.361
633.527	37.558	2.49	0.078	29.365	31.517	7.466	8.631	66.595	116.295
633.679	35.177	2.537	0.067	20.727	22.756	12.693	8.649	66.214	114.447
633.832	34.631	2.525	0.053	16.555	18.516	10.88	8.667	65.697	113.287
633.984	34.683	2.496	0.044	16.572	18.813	7.13	8.631	66.22	114.214
634.136	35.776	2.5	0.039	17.707	20.335	7.431	8.631	65.343	113.231
634.289	35.9	2.515	0.038	19.945	22.987	7.442	8.649	64.712	112.198
634.441	33.98	2.523	0.037	22.356	25.77	7.431	8.649	65.205	112.126
634.594	32.714	2.535	0.036	22.132	25.707	7.428	8.667	65.582	112.167
634.746	34.305	2.543	0.034	19.723	23.114	7.44	8.667	65.728	113.181
634.898	36.001	2.536	0.035	17.722	20.833	8.051	8.649	65.466	113.556
635.051	33.138	2.521	0.038	16.542	19.485	11.17	8.649	65.697	112.567
635.203	28.294	2.505	0.041	16.017	18.834	13.725	8.649	66.336	111.363
635.356	26.112	2.495	0.042	16.073	18.76	11.076	8.631	67.232	111.849
635.508	26.079	2.493	0.041	16.47	19.12	9.64	8.613	67.379	112.077
635.66	25.451	2.491	0.04	16.755	19.408	9.274	8.595	66.626	110.538
635.813	24.267	2.49	0.041	17.075	19.711	8.455	8.577	66.857	110.382
635.965	24.685	2.515	0.045	18.661	21.515	8.986	8.577	67.235	111.198
636.118	30.436	2.522	0.049	22.339	25.653	9.954	8.595	66.332	112.361
636.27	39.09	2.502	0.051	27.434	31.695	9.858	8.595	66.446	116.812
636.422	43.435	2.539	0.053	28.242	33.285	17.006	8.631	67.235	120.487
636.575	42.636	2.584	0.052	24.209	28.855	21.045	8.685	67.351	120.266
636.727	41.039	2.563	0.05	20.527	24.67	14.649	8.685	67.728	120.088
636.88	42.422	2.533	0.046	17.782	21.533	14.747	8.685	67.873	121.083
637.032	43.489	2.541	0.045	16.129	19.616	14.802	8.685	68.018	121.919
637.184	42.05	2.545	0.044	15.211	18.534	14.785	8.667	68.424	121.864
637.337	38.574	2.528	0.044	14.514	17.617	17.973	8.685	67.898	119.095
637.489	40.296	2.518	0.044	14.771	17.746	17.351	8.703	67.119	118.62
637.642	53.32	2.519	0.045	16.49	19.65	10.176	8.685	66.595	124.838
637.794	70.821	2.532	0.051	18.696	22.168	10.988	8.631	66.968	136.685
637.946	82.6	2.579	0.064	19.578	23.019	25.056	8.684	68.278	148.217
638.099	87.079	2.597	0.089	19.433	22.698	41.803	8.81	67.898	151.043
638.251	88.962	2.566	0.106	19.204	22.29	41.699	8.81	67.003	150.622

DEPTH	GR	DENSITY	NEUTRON	RES_DEEP	RES_SHAL	RES_MICR	CAL	DT	DTS
638.404	90.457	2.573	0.096	17.949	20.674	38.937	8.793	66.217	149.723
638.556	86.612	2.584	0.069	16.52	19.03	38.844	8.811	65.814	146.03
638.708	73.989	2.577	0.048	15.153	17.503	32.13	8.793	65.814	136.524
638.861	56.375	2.585	0.04	11.678	13.392	22.318	8.775	65.583	124.715
639.013	41.437	2.592	0.038	8.26	9.435	15.364	8.793	65.35	116.075
639.166	36.243	2.575	0.036	6.773	7.759	12.765	8.811	64.828	112.566
639.318	37.319	2.539	0.037	6.197	7.111	11.868	8.702	64.828	113.092
639.47	38.981	2.483	0.049	5.425	6.259	11.577	8.595	64.945	114.117
639.623	41.377	2.407	0.076	4.221	4.879	11.899	8.613	65.692	116.652
639.775	43.541	2.374	0.104	3.157	3.669	12.135	8.613	66.336	118.931
639.928	42.467	2.379	0.121	2.675	3.132	5.882	8.595	66.48	118.622
640.08	38.186	2.373	0.125	2.661	3.092	2.34	8.613	66.742	116.87
640.232	33.672	2.366	0.128	2.799	3.239	1.959	8.613	66.073	113.469
640.385	31.7	2.376	0.125	2.913	3.37	1.922	8.613	65.176	110.993
640.537	31.43	2.383	0.12	2.954	3.396	2.008	8.631	64.8	110.227
640.69	31.262	2.376	0.11	2.953	3.379	2.099	8.631	64.538	109.704
640.842	31.32	2.37	0.105	2.945	3.339	2.22	8.631	64.798	110.172
640.994	31.709	2.385	0.102	2.884	3.246	2.471	8.631	65.06	110.8
641.147	31.484	2.384	0.099	2.797	3.131	2.936	8.595	64.945	110.498
641.299	30.658	2.383	0.094	2.771	3.066	2.617	8.595	66.353	112.502
641.452	30.755	2.402	0.088	2.804	3.07	2.209	8.631	68.276	115.81
641.604	31.299	2.4	0.091	2.82	3.047	2.138	8.613	69.033	117.363
641.756	30.509	2.399	0.101	2.741	2.916	1.962	8.613	69.294	117.414
641.909	28.366	2.394	0.108	2.597	2.739	2.091	8.631	68.944	115.777
642.061	27.017	2.392	0.107	2.512	2.628	2.347	8.613	68.944	115.128
642.214	27.921	2.403	0.105	2.502	2.615	2.591	8.595	69.555	116.586
642.366	28.983	2.408	0.105	2.5	2.626	2.598	8.613	70.047	117.932
642.518	29.967	2.41	0.108	2.477	2.594	2.367	8.613	70.714	119.548
642.671	31.343	2.379	0.109	2.519	2.597	2.263	8.595	71.121	120.934
642.823	32.179	2.35	0.11	2.589	2.619	2.215	8.595	71.237	121.561
642.976	31.537	2.391	0.113	2.512	2.507	2.135	8.595	71.353	121.428
643.128	29.502	2.415	0.113	2.425	2.376	2.076	8.631	71.469	120.588
643.28	28.015	2.398	0.115	2.412	2.347	2.033	8.649	70.945	118.962
643.433	26.415	2.386	0.117	2.403	2.353	1.947	8.631	69.55	115.849
643.585	26.04	2.392	0.123	2.405	2.357	1.847	8.631	68.161	113.361
643.738	27.304	2.395	0.127	2.415	2.355	1.827	8.631	67.38	112.65
643.89	29.656	2.383	0.123	2.428	2.367	1.864	8.631	66.595	112.438
644.042	30.743	2.371	0.117	2.447	2.374	1.954	8.649	65.96	111.874
644.195	30.326	2.37	0.115	2.396	2.265	2.074	8.649	65.582	111.039
644.347	29.39	2.377	0.119	2.21	2.039	2.181	8.631	64.944	109.527
644.5	29.066	2.376	0.122	1.993	1.814	2.151	8.631	64.944	109.378
644.652	30.203	2.371	0.117	1.891	1.705	2.07	8.631	64.944	109.901
644.804	31.712	2.381	0.11	1.883	1.694	2.057	8.631	66.707	113.607
644.957	31.312	2.4	0.106	1.88	1.687	2.06	8.631	70.551	119.949
645.109	29.549	2.392	0.106	1.879	1.689	2.06	8.613	72.858	122.956
645.262	27.95	2.405	0.104	1.884	1.697	2.056	8.613	76.872	128.865
645.414	27.058	2.438	0.098	1.853	1.667	2.034	8.613	75.526	126.139
645.566	27.11	2.437	0.095	1.708	1.535	1.944	8.613	69.245	115.674
645.719	27.945	2.436	0.097	1.474	1.304	1.772	8.631	67.612	113.341
645.871	28.771	2.434	0.101	1.301	1.128	1.624	8.613	68.046	114.462
646.024	27.974	2.431	0.102	1.204	1.032	1.615	8.613	68.685	115.152
646.176	27.048	2.438	0.101	1.149	0.982	1.657	8.613	69.033	123.704
646.328	26.34	2.417	0.098	1.14	0.973	1.657	8.595	68.654	122.74
646.481	25.81	2.396	0.095	1.142	0.977	1.656	8.613	68.539	122.321
646.633	25.606	2.401	0.093	1.137	0.976	1.656	8.613	68.917	122.913

DEPTH	GR	DENSITY	NEUTRON	RES_DEEP	RES_SHAL	RES_MICR	CAL	DT	DTS
646.786	28.219	2.4	0.09	1.113	0.958	1.695	8.613	68.539	123.296
646.938	32.685	2.387	0.089	1.095	0.932	1.563	8.631	68.308	124.723
647.09	35.118	2.383	0.089	1.154	0.964	1.476	8.613	68.569	126.231
647.243	34.207	2.395	0.089	1.151	0.947	1.403	8.613	68.801	126.268
647.395	30.807	2.409	0.092	1.048	0.854	1.228	8.595	69.033	125.256
647.548	27.907	2.407	0.101	0.994	0.811	1.182	8.595	68.917	123.847
647.7	25.065	2.405	0.112	0.979	0.808	1.179	8.613	68.685	122.282
647.852	22.921	2.403	0.118	0.987	0.809	1.179	8.595	68.917	121.839
648.005	22.451	2.398	0.12	0.989	0.81	1.178	8.595	69.033	121.858
648.157	24.11	2.394	0.119	0.95	0.785	1.166	8.595	68.801	122.106
648.31	26.754	2.378	0.12	0.846	0.691	1.099	8.595	68.801	123.169
648.462	26.965	2.374	0.118	0.801	0.645	1.018	8.577	68.917	123.463
648.614	25.919	2.373	0.117	0.837	0.686	1.091	8.577	68.917	123.039
648.767	25.238	2.374	0.118	0.849	0.702	1.168	8.595	69.408	123.64
648.919	25.709	2.377	0.119	0.839	0.688	1.131	8.613	69.408	123.831
649.072	27.049	2.36	0.116	0.833	0.691	1.109	8.613	69.178	123.964
649.224	28.549	2.353	0.111	0.841	0.69	1.07	8.613	69.294	124.79
649.376	28.627	2.362	0.104	0.838	0.687	1.017	8.631	69.294	124.822
649.529	26.675	2.378	0.102	0.835	0.689	1.153	8.631	69.294	124.02
649.681	25.398	2.391	0.104	0.838	0.688	1.573	8.631	69.149	123.242
649.834	25.165	2.418	0.105	0.831	0.688	1.961	8.613	69.149	123.148
649.986	25.374	2.46	0.104	0.868	0.715	1.858	8.595	69.178	123.284
650.138	24.821	2.485	0.101	0.988	0.812	1.546	8.595	69.062	122.855
650.291	24.692	2.468	0.101	1.02	0.868	1.425	8.613	68.801	122.339
650.443	25.506	2.417	0.102	1.041	0.901	1.395	8.613	68.569	122.252
650.596	25.068	2.406	0.099	1.116	0.967	1.388	8.613	68.569	122.077
650.748	24.615	2.421	0.097	1.119	0.975	1.384	8.613	68.569	121.896
650.9	23.581	2.426	0.098	1.128	0.971	1.371	8.613	68.685	121.69
651.053	22.976	2.425	0.096	1.126	0.977	1.406	8.631	68.917	121.861
651.205	22.774	2.426	0.092	1.119	0.98	2.164	8.631	69.033	121.986
651.358	22.925	2.431	0.086	1.148	1.002	2.784	8.613	69.294	122.507
651.51	23.777	2.459	0.082	1.295	1.131	2.018	8.595	69.555	123.311
651.662	23.509	2.529	0.085	1.612	1.434	1.721	8.595	69.176	122.532
651.815	24.581	2.578	0.091	1.741	1.615	1.754	8.613	68.424	121.624
651.967	27.002	2.533	0.098	1.633	1.557	1.759	8.631	68.134	122.076
652.12	28.932	2.478	0.102	1.604	1.537	1.886	8.631	68.105	122.805
652.272	29.491	2.474	0.102	1.609	1.539	1.797	8.613	67.989	122.823
652.424	28.8	2.472	0.103	1.612	1.541	2.403	8.595	67.728	122.071
652.577	29.471	2.47	0.103	1.628	1.558	5.309	8.595	67.612	122.134
652.729	32.795	2.473	0.103	1.786	1.71	5.396	8.595	67.757	123.762
652.882	37.643	2.479	0.096	2.421	2.342	3.576	8.595	67.757	125.812
653.034	41.577	2.46	0.09	3.552	3.511	3.739	8.613	67.757	127.525
653.186	42.294	2.473	0.088	4.243	4.258	4.553	8.595	68.163	128.607
653.339	42.609	2.527	0.088	4.325	4.334	5.049	8.577	69.318	130.93
653.491	45.926	2.554	0.089	4.337	4.334	5.94	8.595	70.828	135.342
653.644	48.929	2.531	0.086	4.33	4.354	4.424	8.595	70.682	136.503
653.796	45.373	2.496	0.085	4.353	4.39	2.598	8.595	70.682	134.8
653.948	37.001	2.499	0.088	4.154	4.232	2.407	8.595	70.063	129.809
654.101	31.543	2.504	0.09	3.004	3.172	2.428	8.595	68.685	124.932
654.253	32.914	2.505	0.093	2.116	2.326	2.526	8.595	69.176	126.405
654.406	35.819	2.5	0.1	1.868	2.095	2.736	8.595	70.45	130.001
654.558	35.336	2.472	0.105	1.768	1.988	3.881	8.613	71.237	131.239
654.71	33.285	2.47	0.105	1.744	1.964	5.027	8.631	71.468	130.756
654.863	33.013	2.495	0.098	1.736	1.962	3.573	8.631	70.679	129.193
655.015	33.582	2.507	0.093	1.737	1.959	2.355	8.631	69.932	128.073

DEPTH	GR	DENSITY	NEUTRON	RES_DEEP	RES_SHAL	RES_MICR	CAL	DT	DTS
655.168	33.535	2.496	0.085	1.743	1.958	2.31	8.649	69.932	128.053
655.32	39.277	2.493	0.083	1.739	1.967	2.517	8.649	69.932	130.578
655.472	54.047	2.499	0.097	1.725	1.963	2.634	8.649	69.815	137.326
655.625	72.013	2.503	0.124	1.984	2.224	2.819	8.649	68.944	145.04
655.777	83.311	2.53	0.149	3.903	4.189	2.87	8.649	72.042	158.484
655.93	86.014	2.579	0.152	9.3	9.912	2.825	8.649	74.649	166.037
656.082	83.945	2.627	0.139	12.635	13.841	4.334	8.685	72.594	160.11
656.234	80.051	2.638	0.13	12.255	13.462	11.277	8.739	71.468	155.176
656.387	78.528	2.638	0.139	12.233	13.413	12.922	8.739	72.473	156.406
656.539	80.412	2.652	0.16	12.224	13.418	3.005	8.739	69.798	151.769
656.692	83.821	2.648	0.175	12.225	13.446	0.945	8.757	66.22	145.979
656.844	86.215	2.651	0.19	12.23	13.449	0.914	8.81	66.741	148.57
656.996	89.724	2.49	0.234	12.224	13.429	1.148	9.024	65.297	147.473
657.149	93.307	2.355	0.291	12.023	13.307	1.505	9.205	64.559	145.975
657.301	94.173	2.388	0.317	8.773	10.07	1.682	9.626	65.088	147.169
657.454	91.692	2.377	0.299	5.573	6.781	1.483	10.226	63.897	144.478
657.606	88.118	2.346	0.275	4.922	6.21	0.963	10.407	63.234	141.868
657.758	84.661	2.328	0.264	5.206	6.613	0.749	10.425	63.897	141.34
657.911	80.708	2.314	0.269	5.345	6.783	0.476	10.425	65.46	142.505
658.063	76.543	2.298	0.296	5.353	6.77	0.3	10.443	66.481	142.354
658.216	72.806	2.295	0.339	5.35	6.769	0.311	10.443	66.858	141.084
658.368	71.58	2.254	0.378	5.337	6.79	0.938	10.443	67.003	140.721
658.52	72.166	2.144	0.394	5.347	6.816	1.833	10.389	67.003	141.04
658.673	73.663	2.102	0.381	5.559	7.028	0.747	10.495	67.119	142.108
658.825	76.248	2.152	0.367	6.201	7.7	0.38	11.098	66.067	141.303
658.978	78.858	2.295	0.378	6.487	8.075	0.363	10.796	61.76	133.461
659.13	81.809	2.403	0.397	5.601	7.155	0.478	9.921	58.653	128.25
659.282	86.383	2.394	0.376	4.856	6.265	0.585	9.725	58.624	130.589
659.435	89.934	2.395	0.314	4.842	6.265	0.533	9.689	60.027	135.689
659.587	89.916	2.423	0.254	5.039	6.561	1.704	9.49	63.911	144.459
659.74	85.813	2.474	0.219	5.433	7.035	5.423	9.277	70.366	156.381
659.892	82.977	2.512	0.213	6.689	8.583	6.754	9.277	74.687	164.083
660.044	84.186	2.504	0.219	9.4	11.981	9.541	9.241	76.439	168.755
660.197	83.179	2.474	0.195	12.058	15.27	11.043	9.115	73.479	161.56
660.349	75.267	2.472	0.133	13.282	16.889	12.545	9.062	65.302	139.128
660.502	62.968	2.506	0.074	13.825	17.696	6.216	8.99	61.319	124.637
660.654	52.38	2.559	0.053	13.869	17.757	7.26	8.828	59.683	116.692
660.806	47.784	2.592	0.056	13.871	17.743	18.815	8.703	60.192	115.773
660.959	46.37	2.592	0.061	13.891	17.746	19.612	8.685	65.595	125.539
661.111	46.996	2.577	0.065	13.995	17.853	15.226	8.703	68.241	130.89
661.264	48.843	2.568	0.066	13.831	17.502	12.74	8.703	65.594	126.64
661.416	49.156	2.558	0.065	13.737	16.701	11.684	8.685	63.287	122.32
661.568	46.855	2.562	0.06	20.455	23.181	18.541	8.685	61.072	117.083
661.721	42.049	2.558	0.057	40.566	44.734	24.387	8.685	62.084	117.039
661.873	35.986	2.601	0.056	55.781	62.045	22.147	8.685	67.294	124.248
662.026	32.841	2.711	0.056	56.167	62.409	41.052	8.685	69.024	126.097
662.178	34.566	2.747	0.056	55.513	61.573	51.311	8.685	66.377	121.967
662.33	40.057	2.707	0.058	55.572	61.559	31.761	8.703	66.093	123.741
662.483	46.354	2.658	0.075	52.537	57.481	27.752	8.721	65.41	125.177
662.635	54.532	2.664	0.1	32.124	34.036	32.846	8.739	62.854	123.85
662.788	61.468	2.715	0.105	16.039	16.857	26.91	8.792	61.314	123.932
662.94	59.526	2.753	0.084	13.122	14.082	14.221	8.846	60.45	121.31
663.092	50.97	2.75	0.059	16.073	17.432	12.02	8.846	60.334	117.371
663.245	43.547	2.729	0.051	20.442	22.302	53.289	8.846	61.849	117.204
663.397	42.758	2.761	0.065	25.074	27.414	136.415	8.864	62.501	118.115

DEPTH	GR	DENSITY	NEUTRON	RES_DEEP	RES_SHAL	RES_MICR	CAL	DT	DTS
663.55	42.978	2.822	0.1	32.62	35.294	70.827	8.864	61.233	115.807
663.702	46.432	2.73	0.139	24.49	26.146	27.421	8.9	61.233	117.215
663.854	60.562	2.497	0.156	11.764	12.709	10.735	9.008	61.087	123.058
664.007	81.583	2.333	0.17	8.069	8.888	8.285	9.186	60.07	131.231
664.159	101.404	2.301	0.211	6.919	7.632	6.995	9.42	59.551	134.651
664.312	110.747	2.361	0.264	6.284	6.907	3.741	9.564	60.187	136.088
664.464	112.733	2.514	0.288	6.074	6.652	1.031	9.345	61.49	139.034
664.616	114.752	2.627	0.269	5.976	6.543	1.07	9.08	62.132	140.486
664.769	114.34	2.652	0.248	5.916	6.44	3.66	9.044	62.248	140.748
664.921	112.368	2.649	0.235	5.862	6.332	5.731	9.062	63.143	142.773
665.074	104.559	2.583	0.203	6.358	6.785	7.593	9.044	64.449	145.724
665.226	94.888	2.457	0.162	7.981	8.348	3.98	8.99	65.609	148.349
665.378	88.545	2.353	0.145	9.758	9.906	3.559	9.026	66.998	150.577
665.531	89.031	2.373	0.176	9.938	9.8	8.346	9.097	67.989	153.113
665.683	96.937	2.344	0.239	8.964	8.745	2.448	9.187	67.611	152.874
665.836	105.061	2.293	0.301	8.554	8.269	0.454	9.223	67.611	152.874
665.988	108.379	2.37	0.321	8.763	8.333	1.195	9.277	67.989	153.73
666.14	104.656	2.463	0.285	8.956	8.376	7.877	9.259	67.989	153.73
666.293	98.318	2.511	0.215	9.773	9.031	13.546	9.133	68.279	154.385
666.445	91.383	2.541	0.157	9.235	8.602	9.053	9.098	69.059	156.15
666.598	81.948	2.567	0.126	6.574	6.297	5.202	9.098	70.564	154.383
666.75	70.928	2.601	0.11	4.807	4.728	4.334	9.151	71.96	150.752
666.902	63.338	2.586	0.104	4	3.993	3.873	9.133	72.222	147.002
667.055	62.146	2.541	0.102	3.517	3.541	3.967	9.08	71.321	144.522
667.207	61.512	2.537	0.105	3.213	3.28	3.924	9.098	70.046	141.603
667.36	59.188	2.546	0.117	3.137	3.182	2.776	9.062	69.178	138.651
667.512	58.35	2.543	0.131	3.168	3.171	2.073	9.026	68.917	137.703
667.664	59.439	2.52	0.135	3.1	3.102	1.765	9.044	68.801	138.024
667.817	60.28	2.544	0.124	2.611	2.594	2.14	9.044	68.685	138.22
667.969	60	2.671	0.117	2.611	2.524	3.213	8.99	68.569	137.844
668.122	60	2.769	0.136	5.469	5.101	7.037	8.918	68.569	137.844
668.274	60	2.683	0.174	8.831	8.284	19.87	8.9	68.278	137.258
668.426	60	2.584	0.197	7.277	6.976	16.911	8.972	67.612	135.919
668.579	60	2.605	0.186	6.656	6.394	10.44	9.044	67.119	134.929
668.731	60	2.592	0.159	6.632	6.397	4.42	9.151	66.626	133.937
668.884	60	2.563	0.135	6.6	6.395	3.991	9.187	66.481	133.647
669.036	60	2.565	0.124	6.148	5.985	5.524	9.097	66.742	134.171
669.188	60	2.544	0.12	5.07	4.945	3.123	9.062	67.235	135.161
669.341	60	2.529	0.125	3.992	3.893	3.113	9.026	67.351	135.395
669.493	60	2.517	0.131	3.663	3.578	3.732	8.972	67.003	134.696
669.646	60	2.538	0.135	4.01	3.909	4.564	8.918	66.771	134.23
669.798	60	2.595	0.128	4.284	4.168	4.355	8.9	67.262	135.215
669.95	60	2.59	0.118	4.28	4.185	4.987	8.918	67.757	136.211
670.103	60	2.562	0.114	4.275	4.187	7.207	8.918	67.757	136.211
670.255	60	2.569	0.12	4.297	4.175	8.617	8.918	68.393	137.49
670.408	60	2.563	0.133	4.297	4.189	11.226	8.918	68.422	137.549
670.56	60	2.544	0.138	4.326	4.234	10.479	8.918	67.757	136.211
670.712	60	2.552	0.137	4.41	4.351	5.615	8.918	67.118	134.926
670.865	60	2.561	0.125	4.703	4.686	4.608	8.9	67.147	134.984
671.017	60	2.558	0.109	5.334	5.482	6.349	8.864	68.763	138.233
671.17	60	2.583	0.089	5.282	5.614	15.234	8.828	69.291	139.296
671.322	60	2.633	0.076	4.895	5.212	26.585	8.793	67.898	136.494
671.474	60	2.634	0.078	4.891	5.22	21.736	8.81	66.974	134.638
671.627	60	2.571	0.098	4.877	5.221	12.807	8.864	64.498	129.66
671.779	60	2.536	0.127	4.891	5.232	6.599	8.882	61.46	123.553

DEPTH	GR	DENSITY	NEUTRON	RES_DEEP	RES_SHAL	RES_MICR	CAL	DT	DTS
671.932	60	2.566	0.142	4.958	5.301	5.109	8.882	61.575	123.783
672.084	60	2.586	0.141	5.413	5.775	5.075	8.864	62.48	125.603
672.236	60	2.563	0.137	6.309	6.777	7.113	8.846	62.248	125.136
672.389	60	2.55	0.132	6.646	7.24	8.497	8.828	62.392	125.426
672.541	60	2.541	0.126	6.365	6.966	9.771	8.828	62.886	126.419
672.694	60	2.538	0.122	6.251	6.862	9.529	8.828	63.522	127.697
672.846	60	2.554	0.133	6.239	6.872	6.769	8.811	65.195	131.061
672.998	60	2.556	0.151	6.058	6.642	6.641	8.828	67.771	136.24
673.151	60	2.578	0.166	7.214	7.741	6.73	8.828	69.957	140.634
673.303	60	2.592	0.178	10.915	11.557	8.908	8.828	70.715	142.158
673.456	60	2.584	0.193	13.625	14.51	12.266	8.828	70.715	142.158
673.608	60	2.59	0.213	14.415	15.385	12.964	8.828	70.57	141.866
673.76	60	2.624	0.223	15.067	16.062	−999.25	8.846	70.425	141.574
673.913	60	2.641	0.211	14.945	15.978	−999.25	8.846	70.309	141.341
674.065	60	2.611	0.175	14.777	15.835	−999.25	8.846	70.453	141.632
674.218	60	2.594	0.126	14.712	15.741	−999.25	8.828	70.715	142.158
674.37	60	2.593	0.099	14.692	15.697	−999.25	8.811	70.715	142.158
674.522	60	2.599	0.103	14.662	15.687	−999.25	8.846	70.308	141.339
674.675	60	2.603	0.122	13.366	14.299	−999.25	8.9	69.787	140.292
674.827	60	2.6	0.134	9.165	9.891	−999.25	8.9	69.555	139.826
674.98	60	2.566	0.138	10.125	10.82	−999.25	8.882	69.323	139.359

ADDITIONAL DATA FOR FULL EVALUATION

Formation Pressure Data

Depth M (TVD)	Fpress (psia)
624	5177.00
630	5184.30
636	5191.40
642	5198.60
646	5203.60
649	5208.10
652.5	5213.00
662	Tight

Core Description

Depth (m)	Lithology
616–622.5	Shale
622.5–625	Sandstone
625–626.5	Limestone
626.5–637.5	Sandstone
637.5–639	Shale
639–652	Sandstone
652–655.5	Silty sandstone
655.5–660	Shale
660–662	Sandstone
662–664	Limestone
664–675	Shale

Conventional Core Analysis

Depth	Plug Porosity (%)	Horizontal permeability (md)	Grain Density (g/cc)
620	2.0	0.01	2.675
622	2.0	0.02	2.675
624	11.05	22.0	2.665
626	1.0	0.03	2.720
628	9.5	10.5	2.665
630	15.6	135.6	2.662
632	15.0	120.0	2.658
634	7.5	11.0	2.674
636	10.5	15.3	2.666
638	6.0	0.80	2.660
640	17.9	350	2.651
642	15.6	130	2.649

SCAL ANALYSES

Porosity as Fraction at Overburden

Pressure (psi)	Sample 1	Sample 2	Sample 3	Sample 4
50	0.080	0.120	0.140	0.170
500	0.078	0.117	0.137	0.167
1500	0.077	0.115	0.135	0.163
2000	0.076	0.114	0.133	0.161
2500	0.076	0.113	0.132	0.161
4500	0.073	0.110	0.127	0.155
6000	0.072	0.107	0.125	0.153

Brine Permeability in Measured Depth at Overburden

Pressure (psi)	Sample 1	Sample 2	Sample 3	Sample 4
50	2	12	60	540
500	1.757685	10.58051	52.54248	474.8736
1500	1.492423	9.02435	45.0698	402.4928
2000	1.395571	8.429014	42.12812	377.8645
2500	1.318647	7.909027	39.50155	355.9868
4500	1.061196	6.325671	31.62694	286.3587
6000	0.922045	5.504435	27.61859	248.4957

FRF Measurement Data

Porosity	FRF
0.1	89.27854
0.15	44.01849
0.17	33.36747
0.08	133.3982
0.14	43.31214
0.13	58.2223

Resistivity Index Data

Plug 1 (por = 17%)		Plug 2 (por = 15%)		Plug 3 (por = 13%)	
Sw	I = (Rt/Ro)	Sw	I	Sw	I
1	1	1	1	1	1
0.8	1.61592	0.8	1.541192	0.9	1.297253
0.6	2.955611	0.6	2.660774	0.7	2.353045
0.4	8.05591	0.5	3.783241	0.6	3.247888
0.35	10.15069	0.4	6.397332	0.55	3.986601
0.3	12.88598			0.5	5.242788
0.28	15.20306				

Air/Brine Capillary Pressure Curves

		Pc (psi)					
phi	K	3.000	10.000	25.000	50.000	125.000	200.000
0.131	67	0.861	0.617	0.388	0.29	0.239	0.216
0.058	3.7	0.963	0.874	0.782	0.673	0.597	0.525
0.032	3.8	0.942	0.85	0.739	0.638	0.572	0.512
0.179	278	0.73	0.43	0.282	0.214	0.176	0.144

SOLUTIONS TO EXERCISES

CHAPTER 2: QUICKLOOK LOG INTERPRETATION

Exercise 2.1: Quicklook Exercise

	DT	
140	US/FT	40
	NEUTRON	
.45	FRCT	−.15
	DENSITY	
1.95	G/CC	2.95

	CAL	
8	INCH	18
	CR	
0	API	100

"DEPTH"

	RES_DEEP	
.2	OHMM	2000
	RES_SHAL	
.2	OHMM	2000

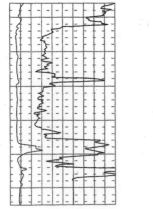

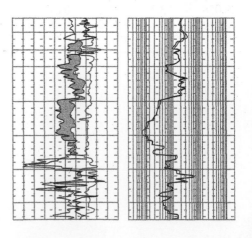

625

650

675

1. $GR_{sa} = 20$, $GR_{sh} = 90$
2. Calculate V_{sh} according to $V_{sh} = (GR - GR_{sa})/(GR_{sh} - GR_{sa})$
3. OWC at 646 m
4. Assume a fluid density of 1.0 g/cc in the water leg and 0.9 in the oil leg.
7. $R_w = 0.02$ ohm
9. OWC at 646 m

11.

Zone	Top (m)	Base (m)	Gross (m)	Net (m)	Av. Por. (m)	S_w
Zone 1	616	622.5	6.5	0		
Zone 2 oil	622.5	646	23.5	21.5	0.108	0.509
Zone 2 water	646	655.5	9.5	9.5	0.124	0.937
Zone 3 water	655.5	675	19.5	1.52	0.05	0.767

12. Suggested stations for pretest measurements are:
    1. 624 m reference density log
    2. 630
    3. 636
    4. 642
    5. 646
    6. 649
    7. 652.5
    8. 662

If the oil type is unknown, it is recommended to take a sample at 630 m using a pump-out module to avoid contamination with WBM.

Exercise 2.2: Using Pressure Data

The following plot may be generated:

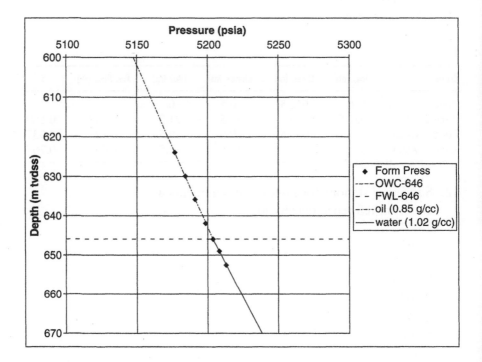

1. From this the FWL is picked at 646 m. Note that because the sand is of reasonable quality no significant gap between the FWL and OWC would be expected. The oil density is 0.85 g/cc. The water density is 1.02 g/cc

2. Oil bearing

3. In order to avoid early water production it is recommended to use the tight streak at 638 m to help avoid early water breakthrough. The following interval is therefore proposed: 622.5–638 m (reference density log). The final evaluated logs should look like this:

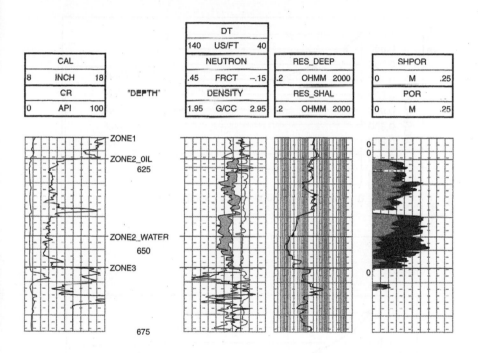

CHAPTER 3: FULL INTERPRETATION

Exercise 3.1: Full Evaluation of the Test1 Well

From the core data it may be seen that the current cutoff point at 50% V_{sh} is appropriate.

Calibrating the log porosity against the core: From the plots of porosity and permeability against isostatic stress the following conversions are estimated:

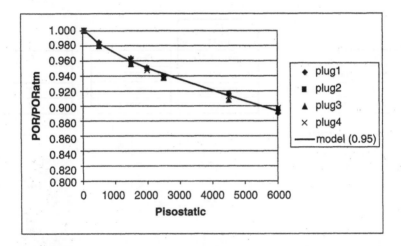

$$\phi_{in\,situ} = \phi_{air} * 0.95$$

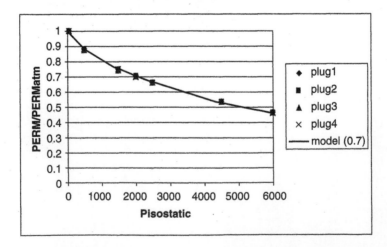

$$k_{in\text{-}situ} = k_{air} * 0.7$$

The poroperm relationship is:

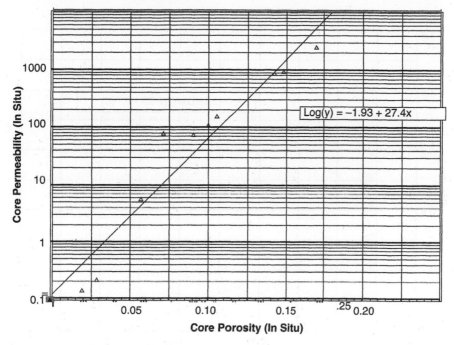

$$k = 10\wedge(-1.93 + 27.4 * \phi)$$

Making a histogram of the core grain densities (excluding the limestone plug) yielded an average grain density of 2.66 g/cc.

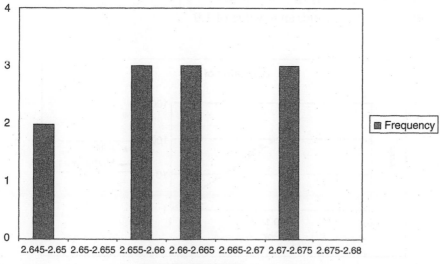

Plotting the in-situ corrected porosities vs the density yields:

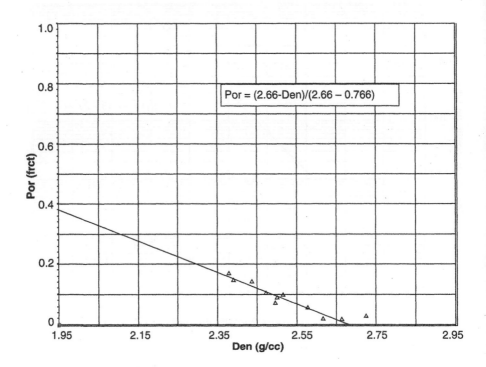

Grain density $= 2.66$ g/cc (fixed by core data)

Fluid density in oil leg $= 0.7663$ g/cc

Plotting F vs. ϕ yields an m value of 1.9

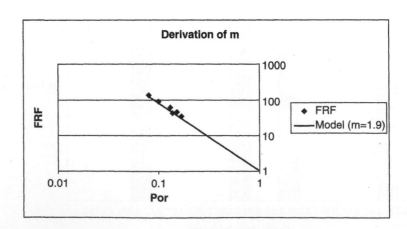

Combining the *I* vs. S_w measurements yields an *n* value of 2.1.

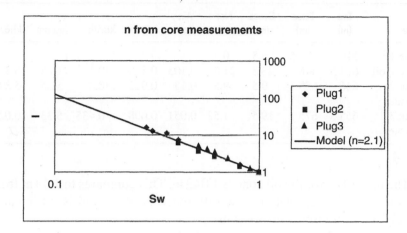

Changing the porosities and *m* required the Pickett plot to be re-performed, yielding a revised value for R_w of 0.025 ohmm.

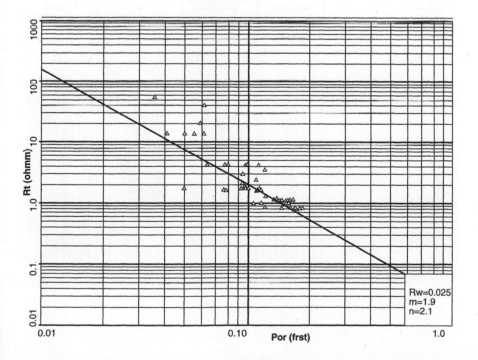

Revised sums and averages are:

Zone	Top (m)	Base (m)	Gross (m)	Net (m)	Av. Por.	Sw	karith	kgeom	kharm
Zone 1	616	622.5	6.5	0					
Zone 2 oil	622.5	646	23.5	21.5	0.105	0.538	40.5	16.5	0.336
Zone 2 water	646	655.5	9.5	9.5	0.13	0.922	188.3	51.5	5.848
Zone 3 water	655.5	675	19.5	1.52	0.051	0.698	0.435	5.33	0.023

The new EHC for the oil zone is 1.043 m. This compares to 1.14 m from the quicklook interpretation. The difference is −9%, and is attributable mainly to the revised value of n used.

The final evaluation of the logs looked like this:

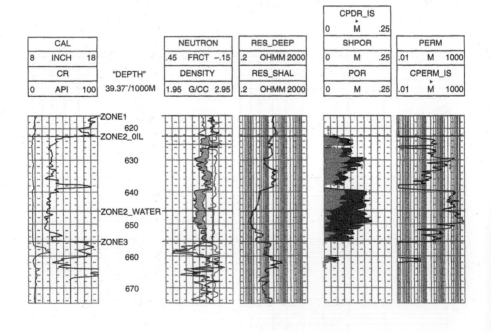

CHAPTER 4: SATURATION/HEIGHT ANALYSIS

Exercise 4.1: Core-Derived *J* Function

A value of S_{wirr} of 0.05 has been assumed. Fitting S_{wr} to J:

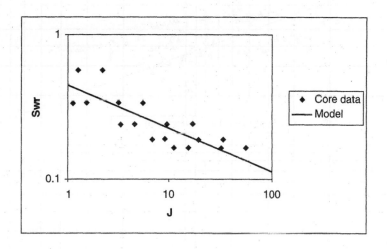

This yields $a = 0.45$, $b = -0.3$. The full function is:

$$S_w = 0.05 + 0.45 * J\wedge(-0.3)$$
$$J = (1.02 - 0.85) * 0.433 * 3.281 * (h) * \sqrt{(k/\phi)/26}$$
$$h = (646 - \text{depth})$$

Where h is in m, k in md, ϕ as fraction.

Using the core derived J function actually resulted in a slight drop (2%) in the equivalent hydrocarbon column. This indicates that thin beds are not much of an issue in this sand, as is evidenced by the logs. Also, the deep resisitivity is a good estimate of R_t.

Exercise 4.2: Log-derived J function

A value of S_{wirr} of 0.05 has been assumed. Fitting S_{wr} to J:

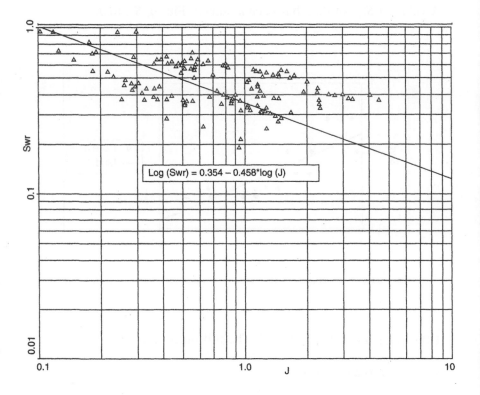

This yields $a = 0.354$, $b = -0.458$. The full function is:

$$S_w = 0.05 + 0.354 * J \wedge (-0.458)$$
$$J = (1.02 - 0.85) * 0.433 * 3.281 * (h) * \sqrt{(k/\phi)/26}$$
$$h = (646 - \text{depth})$$

Where h is in m, k in md, ϕ as fraction.

There is virtually no difference between the core and log derived functions, so it does not really matter which is used. I would nevertheless recommend to use the core-derived function, since this can be more easily updated if further core data becomes available.

Some generic sat/ht curves for a range of porosities are shown below:

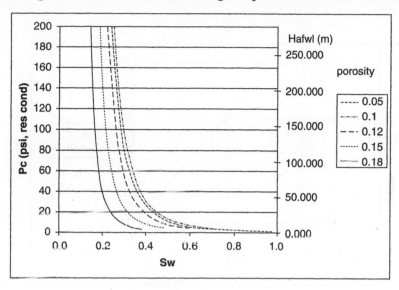

CHAPTER 5: ADVANCED LOG INTERPRETATION TECHNIQUES

Exercise 5.1. Shaly Sand Analysis

The BQ_v relationship is as follows:

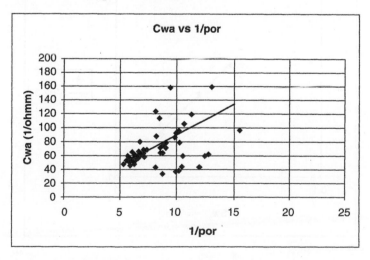

$$BQ_v = (0.18 - \phi)/(0.02 * \phi)$$

Using this function to convert the core F values to F^* yields m^* as follows:

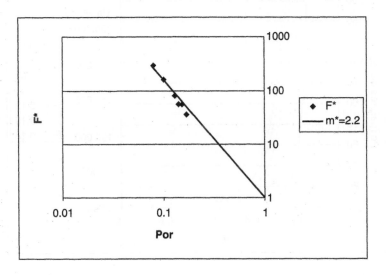

$$m^* = 2.2$$

Likewise, converting I to I^* yields n^* as follows:

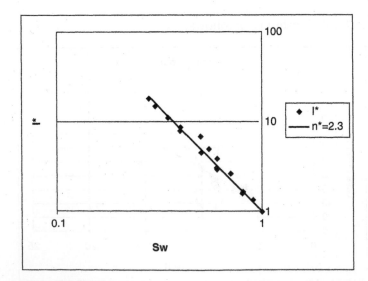

$$n^* = 2.3$$

The new sums and averages are:

Zone	Top (m)	Base (m)	Gross (m)	Net	Av. Por.	Sw
Zone 2 oil	622.5	646	23.5	21.5	0.105	0.480

Using the Waxman-Smits yields slightly more oil (+8%) than Archie, but the effect is small. It is still recommended to use the core-derived J function for STOIIP determination.

Exercise 5.2: Fuzzy Logic

The distributions are as follows:

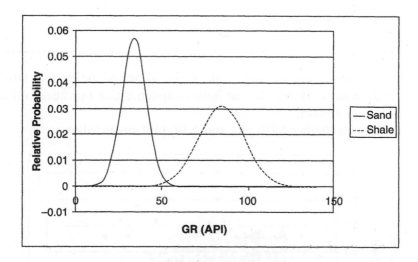

Since this is only a two-variable system, the net/gross will be the same as if a single cutoff was used at the point at which the two distributions intersect (54 API). This contrasts with the previous V_{sh} cutoff used, equivalent to 55 API. The effect of using the fuzzy logic is therefore very little different from applying a normal V_{sh} cutoff.

Exercise 5.3: Thin Beds

The Thomas-Steiber ϕ_d/ϕ_n plot is as follows:

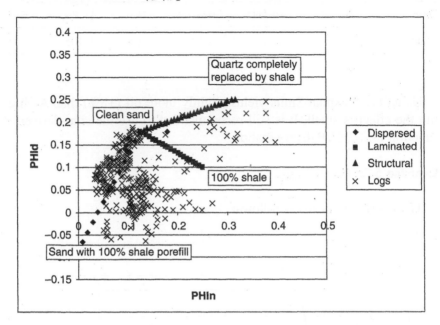

The plot shows that the shale is primarily dispersed, rather than structural or laminated. Hence we can have confidence that our final evaluation is largely correct.

The Thomas-Steiber ϕ_d/V_p plot and ϕ_n/V_p plot are as follows:

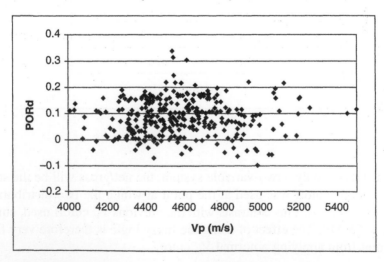

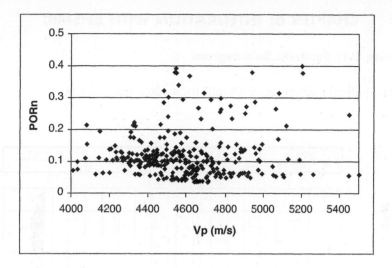

These show considerable scatter, and I do not believe that they show anything useful in this case.

Exercise 5.4: Thermal Decay Neutron Example

The value of S_w is 0.28.

If Σ_{shale} is increased to 30, the value of S_w becomes 0.20. If it is lowered to 20, S_w becomes 0.35. Therefore the uncertainty in S_w resulting from the uncertainty in Σ_{shale} is ±8 s.u.

Exercise 5.5: Error Analysis

Based on the Monte-Carlo analyses, the following ranges are proposed for the average properties for this well:

The exercise was performed on zone 2 only, with the saturation analysis performed only over the oil-bearing part.

Parameter	Mean	s.d.	Suggested Error
Net/gross	0.944	0.0049	±0.01
Porosity	0.108	0.003	±0.006
S_w (Archie)	0.538	0.15	±0.3
S_w (cap curve)	0.532	0.014	±0.03

CHAPTER 6: INTEGRATION WITH SEISMIC

Exercise 6.1: Synthetic Seismogram

The synthetic seismogram should look something like this:

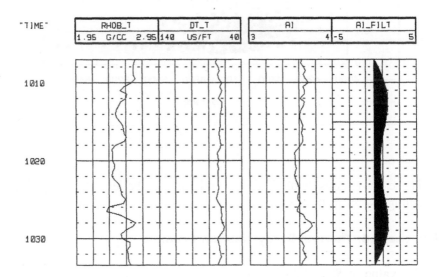

No effects due to the OWC are detectable.

Note that in this field there is very little sonic contrast between the sands and shales.

Exercise 6.2: Fluid Replacement Modelling

The following plot compares AI(oil) with AI(water). As expected, the AI in the water-bearing case is slightly higher than in the oil-bearing case, but the effect is moderate because of the relatively low porosities involved.

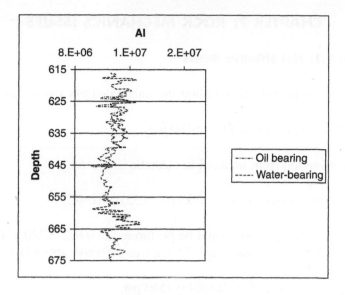

Exercise 6.3: Acoustic Impedance Modelling

The AI distributions for the shales, oil case, and water case are as follows:

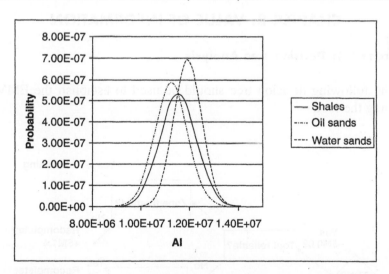

There is virtually no AI contrast between the shales and sands, irrespective of porefill. It may be seen that even if the water sands might be distinguishable from shale (ignoring resolution effects), the effect of the formation becoming oil bearing is that it would seem to disappear to within the shale distribution.

CHAPTER 7: ROCK MECHANICS ISSUES

Exercise 7.1: Net Effective Stress

Ignoring the poroelastic constant the uniaxial stress should be:

$$12,000 \text{ psi} - 12,000 * 0.435 = 6780 \text{ psi.}$$

The conversion from uniaxial to isostatic is given by:

$$\text{Isostatic/uniaxial} = (1 + 0.35)/(3 * (1 - 0.35)) = 0.692$$

Hence the experiments should be performed at $0.692*6780 = 4693$ psi. If the poroelastic constant is 0.85 the true effective stress is given by:

$$12,000 - 0.85 * (12,000 * 0.435) = 7563 \text{ psi.}$$

The experiments should be performed at $0.692*7563 = 5234$ psi.

CHAPTER 8: VALUE OF INFORMATION

Exercise 8.1: Decision Tree Analysis

The following decision tree should be used to establish the EMV of running the tool:

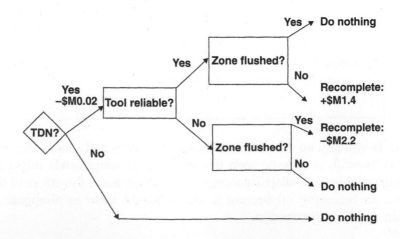

The ΔEMV (in $M) is given by:

$$\Delta EMV = -0.02 + 0.7*0.5*[20*(180,000-60,000)-1]$$
$$+ 0.3*0.5*[20*(-60,000)-1] = \$0.14 \text{ M}$$

The ΔEMV expressed as a function of the reliability, R, is given by:

$$\Delta EMV(R) = -0.02 + R*0.5*(1.4) + (1-R)*0.5*(-2.2)$$

The EMV becomes zero at a value of R of 0.62. So unless the tool will give the right answer at least 62% of the time, it is not worthwhile to run it.

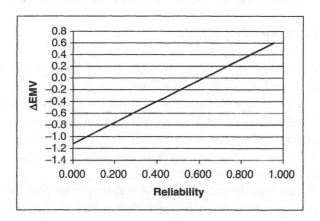

If there is an estimated 70% chance that the zone is not flushed, the ΔEMV becomes:

$$\Delta EMV(R) = -0.02 + R*0.7*(1.4) + (1-R)*0.3*(-2.2)$$

For $R = 0.7$, the ΔEMV becomes $0.468 M; the R required for 0 EMV becomes 0.42

If there is an estimated 30% chance that the zone is not flushed:

$$\Delta EMV(R) = -0.02 + R*0.3*(1.4) + (1-R)*0.7*(-2.2)$$

For $R = 0.7$, the ΔEMV becomes $-0.188 M$; the R required for 0 ΔEMV becomes 0.80.

If there is a 50% chance of the zone being flushed, and the tool predicts the zone to be unflushed with 70% reliability but is only reliable 60% of the time in predicting a flushed case. When the zone is flushed, the EMV becomes:

$$\Delta EMV(R) = -0.02 + 0.70 * 0.5 * (1.4) + (0.4) * 0.5 * (-2.2) = \$0.03 \text{ M}$$

CHAPTER 9: EQUITY DETERMINATION

Exercise 9.1: Optimizing Equity

Note that the test2 well has significantly better sand quality than test1. It is also updip and does not see the OWC.

You would obviously maximize your equity if the determination were to be done on the basis of GBV or NPV. However, you have no technical case to support this, so you must expect that the final basis will be on HCPV. Since neither well encountered gas, there is no gas conversion factor to consider in this example.

The test2 well was drilled with OBM and it appears that the induction tool is saturating, giving unrealistically high values of R_t that in turn lead to very low Archie water saturations.

On this basis you should definitely push for a saturation/height based model for the saturations:

Within the range of reasonable V_{sh} cutoff values that could be chosen there is no dependence between V_{sh} cutoff and equity. This is hardly surprising since the sands are very well defined.

Increasing matrix and fluid densities, both of which have the effect of increasing the porosity, is in your advantage. This is because the test2 oil company gains proportionally less than you do when their porosities are already better than yours.

It also pays you to define a saturation/height function that gives more oil. Hence lowering the a value and making the b value more negative both improve your equity position.

Note that the prize for pushing for the right model could be an improvement in the value of your acreage of about 4% out of 17%, or a 23% improvement. The actual equity involved would change when the full mapping was performed, but it is likely that the improvement will remain roughly constant given that the two wells are indeed representative.

Parameter	Value	Equity	Value	Equity
V_{shco}	0.4	0.199	0.6	0.199
Matrix density	2.65	0.191	2.67	0.206
Fluid density	0.75	0.194	0.95	0.204
A	0.4	0.211	0.5	0.185
B	−0.25	0.206	−0.35	0.192

Base case:

Equity	test1	test2
GBV	0.394	0.606
NPV	0.283	0.717
$HCPV_{arch}$	0.167	0.833
$HCPV_{cap}$	0.199	0.801

CHAPTER 10: PRODUCTION GEOLOGY ISSUES

Exercise 10.1: Dip Magnitude

$$\tan(\alpha) = (100)/(25,000 * .004) = 0.1$$

$\alpha = 5.7$ degrees

Exercise 10.2: Area Depth Graph

The map should look something like this:

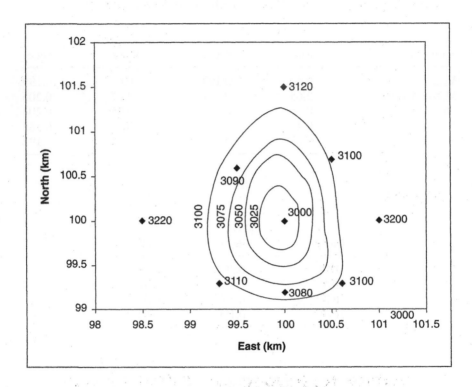

This leads to the approximate area depth relationship:

Depth	Area (m²)
3000	0
3025	15,2305
3050	484,606
3075	896,907
3100	1,66,1509

Plotting this, together with the base of the sand, yields:

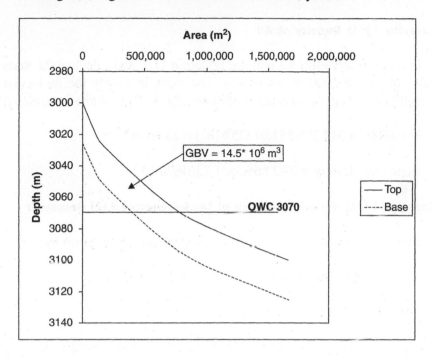

Taking the area above the OWC yields a GBV of $14.5*10^6\,m^3$.

$$NPV = GBV * porosity = 0.2*14.5*10^6 = 2.91*10^6\,m^3$$
$$STOIIP = 2.91*10^6*(0.8)/1.3 = 1.79*10^6\,m^3(=11.26\,MMstb)$$

In order to take into account the change in saturation, the upper part of the sand should be treated separately from the lower half, and the two should be added together. This yields the following elements:

$$GBV1 = 8.06*10^6\,m^3(0-25\,m\ above\ contact)$$
$$GBV2 = 6.44*10^6\,m^3(25\,m + above\ contact)$$
$$STOIIP = (0.2/1.3)*(8.06*10^6*0.7 + 6.44*10^6*0.9)$$
$$= 1.76*10^6\,m^3(=11.07\,MMstb)$$

This is lower than that seen when a single value of saturation is taken, because there is more rock volume lying close to the contact than in the crest of the structure. This illustrates why use of a saturation/height curve is important.

CHAPTER 11: RESERVOIR ENGINEERING ISSUES

Exercise 11.1: Density of Air

In order to solve this we will use Equation 11.3, and calculate the mass and volume of 1000 moles of gas. The mass is simply the molecular weight, i.e., 29 kg. At standard conditions, $Z = 1$. The volume is given by:

$$V = 1000 * 8.31 * 288.5/(1.01325 * 10^5) = 23.66 \, \text{m}^3$$

Hence the density = 29/23.66, or 1.226 kg/m^3

Exercise 11.2: Material Balance of Undersaturated Oil Reservoir

From equation 11.10 the composite compressibility is given by:

$$C = (4.4 * 10^{-5} * 0.2 + 10 * 10^{-5})/0.8 = 1.36 * 10^{-4} \, 1/\text{bar}$$

From Equation 11.11:

$$N_p = [10^7 * (1.3 - 1.25) + 10^7 * 1.3 * 1.36 * 10^{-4} * (300 - 250)]/1.25$$
$$= 4.71 * 10^5 \, \text{m}^3$$

Exercise 11.3: Radial Flow

Just apply Equation 11.19, making sure the units are correct. Hence:

$$\kappa = \phi * \mu * C/k = 0.2 * (10 * 10^{-3}) * (2 * 10^{-4} * 10^{-5})/(0.3 * 10^{-12}) = 13.3$$
$$t = r^2 * (1/4) * \gamma * \kappa * \exp(4 * \pi * k * h * \Delta P/(Q * \mu))$$
$$= (0.15)^2 * (1/4) * (1.781) * (13.3) * \exp[(4 * 3.14 * 0.3 * 10^{-12} * 30) *$$
$$(30 * 10^5)/(200/(24 * 60 * 60) * 10 * 10^{-3})]$$

$$= 306,934 \, \text{secs} \, (3.56 \, \text{days})$$

Exercise 11.4: Horner Analysis

Start by tabulating and plotting P_w vs $\ln[(t + \Delta t)/\Delta t]$:

Time (hr)	Pressure (bar)	Ln[(24 + dt)/dt]	Model [y = 226 −3*(x)]
0	205.32	0	226
0.5	217.89	1.39794	221.8062
1	221.95	1.113943	222.6582
1.5	223.05	0.954243	223.1373
2	223.53	0.845098	223.4647
2.5	223.81	0.763428	223.7097
3	223.96	0.69897	223.9031
4	224.2	0.60206	224.1938
6	224.48	0.477121	224.5686
8	224.68	0.39794	224.8062
10	224.82	0.342423	224.9727
12	224.96	0.30103	225.0969

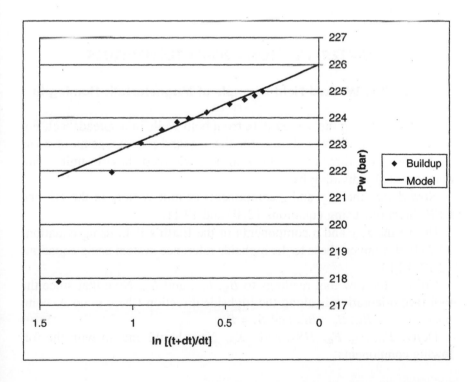

Fitting a straight line through the points corresponding to transient behaviour yields the model line, given by:

$$P_w = 22 - 3.0 * (\ln[(24 + T)/T])$$

Hence the initial reservoir pressure is expected to be 226 bar. Using Equation 11.30, the product $k*h$ can be estimated:

L_1 can be taken as 0, with $P_1 = 226$ bar
L_2 can be taken at 1, with $P_2 = 223$ bar

Hence:

$$k*h = (1/2)*Q*\mu*(L_1 - L_2)/[2*\pi*(P_2 - P_1)]$$
$$= (1/2)*200/(24*60*60)*1.2*1*10^{-3}*(-1)/[2*3.14*-3*10^5]$$
$$= 3.686*10^{-13} \text{m}^2$$

Hence $k = 36.86$ mD.

CHAPTER 12: HOMING-IN TECHNIQUES

Exercise 12.1: Worked Field Example of Magnetostatic Homing In

Start by entering the survey data from both wells in a spreadsheet. For the survey well, derive columns for the x, y and z components of r, the vector along the direction of the well bore. Convert these to unit vector components by dividing by r.

Now derive the x, y and z components of unit vectors in the HS and HSR directions using equations 12.10 and 12.11.

Derive the x, y and z components of the Earth's field using equations 12.7–12.9. Convert these to the highside reference system using equations 12.12–12.14.

Convert the raw tool readings to B_{hs}, B_{hsr}, and B_{ax}. Note that since the tool face orientation is along the highside direction (there is no A_y component), $B_x = B_{hs}$, $B_y = B_{hsr}$ and $B_z = B_{ax}$.

Derive F_{xy}, F_{ax}, F_{tot}, HS_{dir} and AX_{dir}. You should end up with the following components:

Depth m md	F_{xy} µT	F_{ax} µT	F_{tot} µT	HS_{dir} deg	AX_{dir} deg
2700	0.62784	−0.00412	0.62785	349	90.4218
2710	0.77294	0.03687	0.77382	307	87.3133
2720	0.42301	0.14493	0.44715	311	71.1238
2730	0.50055	−0.11071	0.51265	140	102.524
2740	0.5019	−0.0754	0.50753	164	98.5936
2750	0.33035	−0.4646	0.57007	341	144.659
2760	0.54611	−1.8051	1.8859	274	163.25
2765	0.99398	−3.041	3.19932	271	161.982
2770	2.83984	−5.5069	6.19602	243	152.798
2772.5	6.2726	−6.72687	9.19762	245	137.071
2775	11.1798	−7.94685	13.7164	241	125.47
2777.5	17.907	−6.11682	18.9229	243	108.915
2780	23.7569	0.5832	23.7641	241	88.6387
2782.5	23.9532	6.64478	24.8577	241	74.5334
2785	18.1464	12.3364	21.9426	233	55.8195
2787.5	11.919	11.4479	16.5263	213	46.1783
2790	8.68332	8.9695	12.4841	188	44.0936
2792.5	7.39378	4.7118	8.7675	169	57.5212
2795	5.40196	2.8941	6.12838	165	61.8512
2797.5	4.76053	1.3264	4.94186	164	74.4686
2800	3.47112	0.4787	3.50397	159	82.1896
2805	2.24769	0.11865	2.25082	146	87.0224
2810	1.71776	0.1186	1.72185	133	86.094
2820	2.33087	−0.3621	2.35883	39	98.8804

The field pattern is characteristic of a south monopole. Firstly it is noted that the maximum in F_{tot} occurs at 2781.5 ft in the survey well. This must be the point of closest approach between the survey well and pole.

Using the formula for deriving the pole distance from the observed half-width of F_{tot} (equation 12.18), the distance to the pole is found to be 7.5 ft. Using the formula for converting HS_{dir} into an inclination with respect to the horizontal (Equation 12.22) we find that ϕ is 24.2 degrees. The true vertical depth of the pole is therefore:

$$2130.6 - 7.5 * \sin(24.2) = 2127.5 \text{ ft ss.}$$

Where 2130.6 is the true vertical depth at 2781.5 ft md in the relief well. The pole strength may be derived from the formula:

$$F_{tot} = P/(4 * \pi * x^2)$$

Using $F_{tot} = 25.6\,\mu T$, $x = 7.5\,ft$ yields $P = 1681\,\mu Wb$.

The field due to such a monopole may be modelled and compared to the measured readings. In order to do this, construct a spreadsheet as follows:

For all the survey points model the field that would be seen in the high-side reference system:

Determine the x, y and z components of unit vectors in the HS, HSR, and RVEC directions. Calculate $\underline{F}.\underline{HS}\wedge$, $\underline{F}.\underline{HSR}\wedge$ and $\underline{F}.\underline{RVEC}\wedge$, in order to determine F_{hs}, F_{hsr} and F_{ax}. Calculate AX_{dir} and HS_{dir} from these quantities.

Overlay the modelled and measured data. Initially the match is not very good. Because of the nature of the well trajectories (relief well passing to the right of the target well going north to south with an inclination of ~50 deg) the peaks can be aligned by moving the target well north or south. The height of the peaks may be adjusted by moving the target well east or west. By manually adjusting the assumed pole position on a trial-and-error basis it was found that the best fit occurred when the target well was moved 1.5 ft west and 8.5 ft north. After shifting the target well you should end up with the following components:

Depth	TVD	F_{hs}	F_{hsr}	F_{ax}	F_{tot}	F_{xy}	AX_{dir}	HS_{dir}
2685.00	2073.40	−0.01	−0.01	−0.16	0.16	0.01	175.14	227.71
2697.00	2080.70	−0.01	−0.02	−0.20	0.20	0.02	174.77	234.06
2707.00	2086.70	−0.02	−0.01	−0.26	0.26	0.02	175.29	220.24
2717.00	2092.70	−0.03	−0.03	−0.35	0.35	0.04	173.16	230.48
2727.00	2098.70	−0.02	−0.06	−0.48	0.49	0.06	172.93	254.70
2740.00	2105.70	−0.08	−0.11	−0.80	0.81	0.14	170.54	234.72
2750.00	2111.70	−0.16	−0.26	−1.35	1.39	0.30	167.40	238.53
2760.00	2117.60	−0.52	−0.75	−2.67	2.82	0.91	161.22	235.38
2770.00	2123.70	−2.13	−3.83	−6.52	7.85	4.38	146.16	240.89
2780.00	2129.70	−12.92	−20.63	−4.66	24.78	24.34	100.88	237.97
2790.00	2135.80	−3.93	−6.29	8.35	11.17	7.42	41.63	238.02
2800.00	2141.90	−0.76	−1.19	3.32	3.61	1.41	22.95	237.43
2810.00	2148.00	−0.22	−0.35	1.60	1.65	0.42	14.63	238.16
2820.00	2154.10	−0.02	−0.15	0.92	0.93	0.15	9.38	262.97
2830.00	2161.20	−0.05	−0.07	0.58	0.58	0.09	8.99	234.87
2840.00	2167.30	−0.03	−0.05	0.40	0.41	0.06	7.91	237.10
2850.00	2173.50	−0.02	−0.03	0.30	0.30	0.03	6.10	237.94
2860.00	2179.70	−0.01	−0.02	0.23	0.23	0.02	5.33	237.96

The Field components should look as follows:

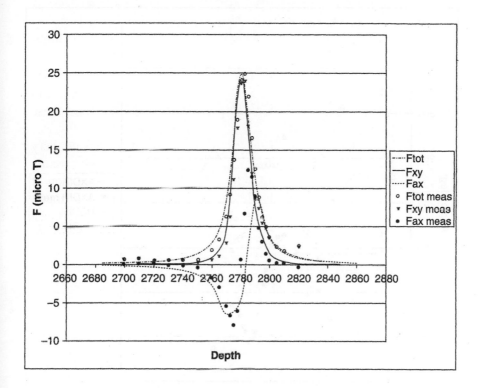

It is seen that for AX_{dir} and HS_{dir}, good agreement is seen only in the vicinity of the pole. This is because the directions become poorly defined once the fields become weak far from the pole.

The asymmetric behavior below 2781.5 ft may be attributed to a second, weaker pole along the target well axis. It may be seen that the addition of a second, north pole at 2147.5 ft in the target well, having a strength of 1000 μWb, could improve the match.

Exercise 12.2: Interpretation of Electromagnetic Homing-In Data

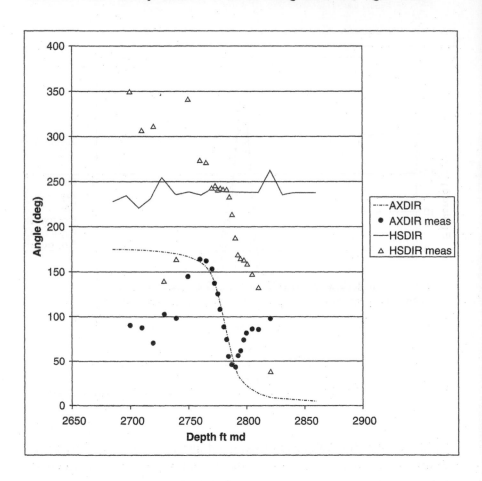

From plotting the HS_{dir} and H curve versus depth it is apparent that the point of closest approach occurs at 2795 ft md.

Taking the width of the HS_{dir} curve for −45 to +45 degrees yields $\Delta = 12$ ft. Hence $D = 12/2*\tan(50) = 7.1$ ft from equation 12.33.

Taking the width of the Intensity curve at half the maximum, and subtracting a background signal of $52\,\mu A/m/A$ yields $\Delta = 37$ ft. This yields a distance of 5.0 ft from equation 12.34.

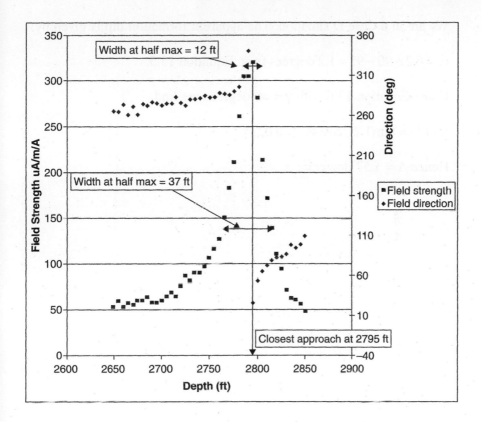

The difference cannot be explained, although it may be a combination of tool accuracy, noise, anisotropy, and variations in target current. Since most of these factors will affect the two AC magnetometers equally, probably the range determined from analyzing HS_{dir} is the more accurate of the two.

CHAPTER 13: WELL DEVIATION, SURVEYING, AND GEOSTEERING

Exercise 13.1: Formation Dip from Up/Down Logs

Using Equation 13.1 we see that the relative dip is:

$$\theta = \arctan[(8.5/12)*(1/3.281)/2] = 6.2 \text{ degrees}$$

We are in a Case D situation. The apparent formation dip is given by:

$\alpha = 6.2 + 90 - 95 = 1.2$ degrees from Equation 13.5

Using Equation 13.6 with $\gamma = 40$ degrees we find:

$\tan(\Delta) = \tan(1.2)/\cos(40) = 0.027$

Hence $\Delta = 1.57$ degrees.

ADDITIONAL MATHEMATICS THEORY

For readers who do not have a mathematics, engineering, or physics degree, some of the basic mathematical principles assumed in this book may be problematic. Therefore, this Appendix is designed to provide a fuller explanation of some of the theoretical derivations used in the chapters.

A4.1 CALCULUS

Differentiation is the taking of the gradient of a function with respect to one of the input variables. Start by considering the function:

$$y = a * x + b.$$

This is the equation of a straight line having a gradient of a and intercept on the y axis at b.

The differential of y with respect to x is a function that describes the rate of change of y with x. It is denoted by dy/dx, where d represents the infinitesimally small increments of y and x. For the function given:

$$dy/dx = a.$$

For most functions that engineers encounter, the differentials are simply known by heart, or can be looked up in mathematical handbooks. Table A4.1 gives most of the functions one is likely to come across:

It is also possible to take the differential of dy/dx, in which case the result is referred to as d^2y/dx^2 or $d/dx(dy/dx)$. Where a function depends

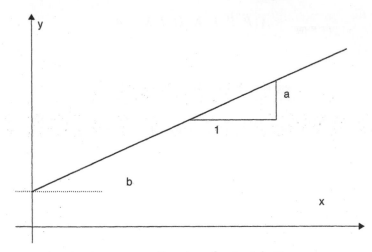

Figure A4.1 Equation of a Straight Line

Table A4.1

Function	dy/dx
$y = x^n + a \ (n \neq 0)$	$n * x^{n-1}$
$y = e^x$	e^x
$y = \log(x)$ (log is natural logarithm base e)	$1/x$
$y = \sin(x)$ (x must be in radians = deg*π/180)	$\cos(x)$
$y = \cos(x)$ (x must be in radians = deg*π/180)	$-\sin(x)$
$y = \tan(x)$	$\sec^2(x)$ (sec = 1/cos)
$y = a^x$	$a^x * \log(a)$

on more than one input variable, the situation is a bit more complex. Consider the function:

$$t = a * x + b * y$$

In order to derive *dt/dx* you also need to know how *y* will vary with *x*, if at all. In most engineering applications, *x* and *y* might be parameters such as pressure or temperature, which one can control in a laboratory. A special notation convention is used when the differential with respect to one variable is derived while keeping the other variables constant. Hence the partial differential of *t* with respect to *x* while keeping *y* constant is denoted as ∂*t*/∂*x*, or sometimes ∂*t*/∂*x*|$_y$.

For the function given $\partial t/\partial x = a$, the constant terms becoming zero on differentiation. Integration is just the opposite of differentiation. While taking the differential of a function of one variable yields the gradient of a graph of y vs. x, integrating the function yields the area under the graph (from the curve to the $y = 0$ axis).

Consider again the function $y = a*x + b$. The integral of y with respect to x is denoted by:

$$\int y\,dx = \int (a*x+b)\,dx = 0.5*a*x^2 + bx + c$$

where c is a constant and the $\int$ is like a drawn-out S, indicating that the summation is made over infinitessimally small increments of dx. Since integration is the opposite of differentiation:

$$\int (dy/dx)\,dx = y + c.$$

The constant c arises because the gradient (dy/dx) contains no information about any fixed offset of y from the $y = 0$ axis (which disappears during differentiation).

In order to determine the area under a graph of y vs. x, one needs to specify a start and end point for x. These are placed at the bottom and top of the $\int$ sign. The integral becomes a definite integral. As with differentiation, most engineers have committed to memory the common indefinite integrals they are likely to need. Table A4.2 shows those commonly used.

For a definite integral, the normal procedure is to first evaluate the indefinite integral, then subtract the value of the function at the start value from that at the end value to get the area under the graph. Hence, for our

Table A4.2

Function	$\int y\,dx$
$y = x^n + a\ (n \neq -1)$	$(1/n + 1)*x^{n+1} + a*x + c$
$y = e^x$	$e^x + c$
$y = \log(x)$ (log is natural logarithm base e)	$x*(\log(x) - 1) + c$
$y = 1/x$	$\log(x) + c$
$y = \sin(x)$ (x must be in radians $= \deg*\pi/180$)	$-\cos(x) + c$
$y = \cos(x)$ (x must be in radians $= \deg*\pi/180$)	$\sin(x) + c$
$y = \tan(x)$	$-\log(\cos(x)) + c$
$y = a^x$	$a^x/\log(a) + c$

example function $y = a*x + b$ evaluated between x_1 and x_2, the integral becomes:

$$\int_{x_1}^{x_2} (a*x+b)dx = [0.5*a*x^2 + bx + c] = 0.5*a*(x_2^2 - x_1^2) + b*(x_2 - x_1)$$

In many real engineering problems, data are presented as sampled at discrete intervals (e.g., 0.5-ft sampling increment for logs) and cannot be described by simple mathematical functions. For these data, a numerical differentiation or integration may also be performed without resort to calculus.

Say, for instance, one wanted to make a differential of a GR log with respect to depth. The procedure would be to simply take the difference between successive data values at each increment and divide by the depth increment. Taking the integral would involve just adding successive data values multiplied by the depth increment.

A4.2 SPECTRAL (FOURIER) ANALYSIS

For any wireline log sampled in depth, it is possible to think of the log as being composed of a complex mixture of cosine waves that, when added together in the right proportions, yield the log. The cosine functions will have the form:

$$y_i = A_i * \cos(2 * \pi * x / \lambda_i + \phi_i)$$

where

A_i = the amplitude of the component i
$(1/\lambda_i)$, or k_i, = the wavenumber of the component i
ϕ_i = the phase of the component i.
If $L(x)$ is the complete log, we can say:

$$L(x) = \Sigma yi.$$

Spectral analysis is the mathematical determination of the set of A_i and ϕ_i as a function of k_i. The determination of the spectra is performed using computer algorithms, which will not be discussed here.

The range of k_i that needs to be used is 0 (corresponding to a cosine wave of infinite wavelength) to 1/sample increment (since variations at a smaller scale than the sampling increment cannot be detected anyway).

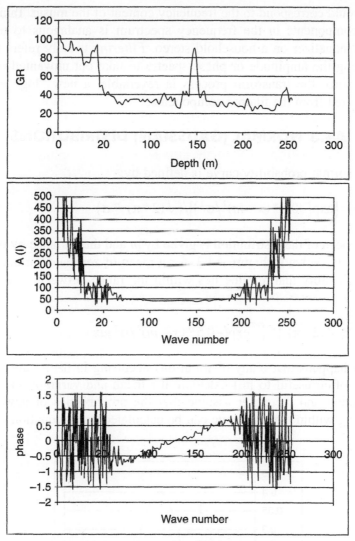

Figure A4.2 Example of Spectral Analysis

Hence a single log may be broken down into an amplitude and phase spectrum, which together give you the relative proportions and phase of each cosine wave as a function of wavenumber needed to add up to the log. This is illustrated in Figure A4.2.

In the event that the log data are sampled in time rather than depth, the spectra are a function of frequency rather than wavenumber. If the log were in fact a sound wave, such as a piece of music, the amplitude spectrum would correspond to the frequency content of the music. Taking out some components in the frequency spectrum is analagous to using a graphic equalizer on a household stereo. *Filtering* usually refers to manipulating the amplitude or phase spectra to take out unwanted components. Since the transform process is reversible, a new "log" may be constructed from the filtered components.

A4.3 NORMAL (GAUSSIAN) DISTRIBUTIONS

The normal probability curve is defined by:

$$p(x) = \left[\exp(-0.5 * (x - m)^2 / \sigma^2)\right] / (\sqrt{(2 * \pi)} * \sigma) \tag{A4.1}$$

The mean of the distribution is given by m and the variance by σ^2. This function is shown in Figure A4.3, for a mean of zero and variance of 1. The probability that a value lies within the range between x_1 and x_2 is given by:

$$p = (2/\sqrt{(2 * \pi)}) * \int_{x_1}^{x_2} \exp(-0.5 * (x - m)^2 / \sigma^2) dx \tag{A4.2}$$

It so happens that many distributions occurring in nature are normal, so it is often useful to just calculate the mean and variation of a given distribution and thereafter assume that the probabilities of new values occurring within a given range can be calculated using equation A4.2.

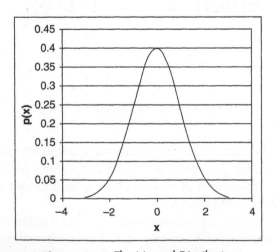

Figure A4.3 The Normal Distribution

In particular, it is worth noting that 68.3% of values will fall within 1 standard deviation (*SD* is the square root of the variance), 95.4% within 2 *SD* and 99.7% within 3 *SD* of the mean.

A4.4 VECTOR MECHANICS

An understanding of the basics of vector mechanics is essential to being able to deal with components in the highside refence system, as discussed in Chapter 12. Consider a cartesian reference system defined by three orthogonal axes: *x*, *y*, and *z*. A vector is essentially just a way of writing directions, in terms of how far you have to travel in the *x*, *y*, and *z* directions to get from one point to another in space. Hence the vector linking the origin to the point A located at (a_1, a_2, a_3) is denoted by $\underline{a}$, and has the components:

$$\begin{pmatrix} a1 \\ a2 \\ a3 \end{pmatrix}$$

The length of the vector, denoted by *a*, is given by $\sqrt{(a_1^2 + a_2^2 + a_3^2)}$. A unit vector is one that has a length of 1. To convert a to a unit vector, each component would have to be divided by $\sqrt{(a_1^2 + a_2^2 + a_3^2)}$ and the vector would then be designated by $\hat{\underline{a}}$. Taking the **scalar product** (sometimes called the **dot product**) of two vectors allows the angle between them to be determined:

$$\underline{a} \cdot \underline{c} = a * c * \cos(\theta) = (a_1 * c_1 + a_2 * c_2 + a_3 * c_3) \qquad (A4.3)$$

where *a*, *c* are the magnitude of *a* and *c*, and θ is the angle between the vectors. The **vector product** of two vectors (sometimes called the **cross product**) generates a new vector that is orthogonal (i.e., at right angles) to both. Hence:

$$\begin{pmatrix} a1 \\ a2 \\ a3 \end{pmatrix} \wedge \begin{pmatrix} c1 \\ c2 \\ c3 \end{pmatrix} = \begin{pmatrix} a2 * c3 - a3 * c2 \\ a3 * c1 - a1 * c3 \\ a1 * c2 - a2 * c1 \end{pmatrix} \qquad (A4.4)$$

The vector defined by $\underline{a} \wedge \underline{c}$ has magnitude $a * c * \sin(\theta)$. The direction of the vector product will follow a right-hand corkscrew rule as one goes from **a** to **c**. Hence, say **a** lies in a northerly direction and **c** in an easterly direction; the cross-product would point downward into the Earth. Of particular reference to surveying is the fact that taking the vector product of a unit gravity vector with a vector in the hole direction yields a horizontal vector 90 degrees to the right of highside in the sensor plane. Taking the vector product of this vector with a vector in the hole direction yields a vector in the direction of highside.

A4.5 PROBABILITY THEORY

It may be helpful for readers who are involved with VOI calculations to have more background information on probability theory to understand better the concepts of EMV and reliability. Some of these will be explained in this section. Suppose that you have discovered an oil field. This field has many uncertainties surrounding it. However, you have determined a field development plan that you intend to carry out, and wish to know the EMV of such a plan.

In reality, depending on the actual true nature of the field, your plan may be either a very good one, a very bad one, or somewhere in between. While the true nature of the field may obviously have an infinite number of different states, consider for now that there are N possible states that more or less encompass all the range of possibilities.

For a particular state i (out of the N possibilities), your field development plan will yield a value of NPV(i), this being the present value of the net revenue minus expenditure over the life of the field. The state i has a probability P(i) of being close to the true state of the field. Clearly it must be true that

$$\sum_{i=1}^{N} P(i) = 1.0 \qquad\qquad (A4.5)$$

The EMV will be given by:

$$EMV = \sum_{i=1}^{N} (i) * NPV(i) \qquad\qquad (A4.6)$$

Some of the NPV(i) may be negative (e.g., if the field is much smaller than originally thought), and some may be very positive (if the field is larger than expected). The final EMV should certainly be positive, or else the whole development would not be worth embarking on.

Now consider that someone proposes an amendment to the field development plan. An example might be the addition of a data acquisition program such as thermal decay logging. This program will not, of course, change $P(i)$, but it will change $NPV(i)$. More money will be spent (reducing $NPV(i)$) but in return, if the data are reliable and useful (which they may be for only some of all possible field states), there will result an increase in revenue or decrease in other costs, having the net effect of increasing $NPV(i)$.

Since the $NPV(i)$ has changed, the EMV will change, to EMV'. The EMV of the proposed change, which we will call ΔEMV is given by:

$$\Delta EMV = EMV' - EMV \tag{A4.7}$$

It is important to note that ΔEMV, irrespective of any issues concerning reliability, depends on all the possible states of the field, not just the base case.

In order to introduce the concept of reliability, it will be much simpler to consider for now that the field has only two possible states, which will be denoted as S_1 and S_2. The EMV of the field is then approximated by:

$$EMV = P(S_1) * NPV(S_1) + P(S_2) * NPV(S_2). \tag{A4.8}$$

Now consider a proposal for a change to the FDP. This will involve the acquisition of data at a cost of Z and be such that a parameter C will be determined as being true or false. C is such that:

1. If C is true, the field is known to be definitely in state 1. If C is false, the field is known to be definitely in state 2. Knowing which state the field is in would allow the FDP to be optimized.
2. If C is true, the FDP may be optimized to yield a new NPV given by $NPV(S_1$ and $C)$. the field being in state S_1.
3. Likewise if C is false, the FDP may be optimized to yield a new NPV given by $NPV(S_2$ and $C')$, the field being in state 2.

The change in the EMV is given by:

$$\Delta EMV = -Z + P(S_1) * NPV(S_1 \text{ and } C)$$
$$+ P(S_2) * NPV(S_2 \text{ and } C') - EMV. \tag{A4.9}$$

For the change to be worthwhile, we clearly require that at least one of the $NPV(S_1$ and C) or $NPV(S_2$ and $C')$ be greater than $NPV(S_1)$ or $NPV(S_2)$.

The VOI is clearly given by VOI $= (\Delta EMV - Z)$. Now consider the effect of reliability. This means in effect that sometimes the parameter C will be found to be true even though the field is in state 2, and vice versa. The reliability is expressed via:

$$R = P(C/S_1)$$

i.e., the probability that C is found to be true when the field is indeed in state 1. For simplicity we will also assume that this is the same as $P(C'/S_2)$, i.e., the probability that C is found to be false when the field is indeed in state 2. The introduction of R clearly leads us to have to consider the additional NPV scenarios:

1. NPV(S_1 and C'): the NPV realized when the field is thought to be in state 2 but is actually in state 1.
2. NPV(S_2 and C): the NPV realized when the field is thought to be in state 1 but is actually in state 2.

We will now calculate ΔEMV, introducing these additional scenarios:

$$\Delta EMV = -Z + P(S_1)*[P(C/S_1)*NPV(S_1 \text{ and } C) + P(C'/S_1)*$$
$$NPV(S_1 \text{ and } C')] + P(S_2)*[P(C'/S_2)*$$
$$NPV(S_2 \text{ and } C') + P(C/S_2)*NPV(S_2 \text{ and } C)] - EMV \quad (A4.10)$$

Using our definition of R:

$$\Delta EMV = -Z + P(S_1)*[R*NPV(S_1 \text{ and } C) + (1-R)*$$
$$NPV(S_1 \text{ and } C')] + P(S_2)*[R*NPV(S_2 \text{ and } C') +$$
$$(1-R)*NPV(S_2 \text{ and } C)] - EMV \quad (A4.11)$$

Another way to look at this is that we have four possible combinations of C, C', S_1, and S_2, each having a probability and NPV associated with it. The EMV is therefore:

$$\Delta EMV = P(S_1 \text{ and } C)*NPV(S_1 \text{ and } C)$$
$$+ P(S_2 \text{ and } C)*NPV(S_2 \text{ and } C)$$
$$+ P(S_1 \text{ and } C')*NPC(S_1 \text{ and } C')$$
$$+ P(S_2 \text{ and } C')*NPV(S_2 \text{ and } C')$$
$$- Z - EMV \quad (A4.12)$$

Also

$$P(S_1 \text{ and } C) + P(S_2 \text{ and } C') + P(S_1 \text{ and } C') + P(S_2 \text{ and } C') = 1 \qquad (A4.13)$$

Now $P(C \text{ and } S_1)$, the probability of both C and S_1 occurring, is given by:

$$P(C \text{ and } S_1) + P(C/S_1) * P(S_1) = R * P(S_1)$$

and so on for the other combinations. Replacing $P(C \text{ and } S_1)$ and similar terms in equation A4.12 yields back the same result as equation A4.7. It is often useful to make a plot of ΔEMV as a function of R. In this way it is possible to determine the value of R for which a particular data acquisition campaign becomes viable.

The above concepts can obviously be extended to cover more than two states, and specialized software is available that will allow the EMV to be calculated relatively simply. Note that the reliability of the tool has been defined as $P(C/S_1)$ etc. One is also interested to know $P(S_1/C)$, i.e., the probability of the field being in state 1 given that the tool yields a result C. To make this conversion it is necessary to use Bayes' theorem. This uses the fact that:

$$P(C \text{ and } S_1) = P(C/S_1) * P(S_1) = P(S_1 \text{ and } C) = P(S_1/C) * P(C) \qquad (A4.14)$$

$$P(C) = P(C/S_1) * P(S_1) + P(C/S_2) * P(S_2). \qquad (A4.15)$$

Combining these equations:

$$P(S_1/C) = [P(S_1) * P(C/S_1)]/[P(S_1) * P(C/S_1) + P(S_2) * P(C/S_2)] \qquad (A4.16)$$

$$P(S_1) * R/[P(S_1) * R + P(S_2) * (1 - R)], \text{ since } R = P(C/S_1) \qquad (A4.17)$$

Likewise:

$$P(S_1/C') = [P(S_1) * P(C'/S_1)]/[P(S_1) * P(C'/S_1) + P(S_2) * P(C'/S_2)] \qquad (A4.18)$$

$$P(S_1) * (1 - R)/[P(S_1) * (1 - R) + P(S_2) * R] \qquad (A4.19)$$

$$P(S_2/C) = [P(S_2) * P(C/S_2)]/[P(S_2) * P(C/S_2) + P(S_1) * P(C/S_1)] \qquad (A4.20)$$

$$P(S_1) * (1 - R)/[P(S_1) * (1 - R) + P(S_2) * R] \qquad (A4.21)$$

$$P(S_2/C') = [P(S_2) * P(C'/S_2)]/[P(S_2) * P(C'/S_2) + P(S_1) * P(C'/S_1)] \quad (A4.22)$$

$$P(S_2) * R/[P(S_2) * R + P(S_1) * (1 - R)], \text{ since } R = P(C'/S_2). \quad\quad (A4.23)$$

Note that it is always true that:

$$P(S_1/C) + P(S_2/C) = 1 \quad \text{and} \quad P(S_1/C') + P(S_2/C') = 1.$$

Note also that only in the special case that:

$$P(S_1) = P(S_2) = 0.5$$

is it true that:

$$P(S_1/C) = P(C/S_1)$$

A4.6 LEAST SQUARES FIT AND CORRELATION

Consider a series of points $(x_1, y_2) \dots (x_n, y_n)$. When plotted these points may lie approximately on a straight line or a curve or be scattered randomly. **Least squares fit** is a way to find the coefficients of a function approximating the behavior of the data.

Consider data which may be approximated by the formula $y = a*x + b$. The line yielded by this equation is known as the line of regression of y on x. Forming the sum of the squares of all the deviations of the given numerical value of y from the theoretical values:

$$S = \Sigma(y - a*x - b)^2 \quad\quad\quad\quad (A4.24)$$

S is minimized when $\partial S/\partial a = \partial S/\partial b = 0$. This occurs when:

$$\Sigma_2 * x * (y - a*x - b) = 0 \quad \text{and} \quad \Sigma_2 * y * (y - a*x - b) = 0. \quad (A4.25)$$

Solving these equations:

$$a = [n * \Sigma x * y - \Sigma x * \Sigma y]/[n * \Sigma x^2 - (\Sigma x)^2] \quad\quad\quad (A4.26)$$

$$b = [\Sigma y * \Sigma x^2 - \Sigma x * \Sigma x * y]/[n * \Sigma x^2 - (\Sigma x)^2] \quad\quad\quad (A4.27)$$

where n is the number of samples.

The regression of y on x assumes that the x values in the data are always correct and that the scatter occurs in the y variable. Similarly, the line of regression of x on y may be derived simply by first setting $x = (1/a)*y + (-b/a)$ and using equations 13.24–13.27 in an identical way.

A set of points on a plane may exhibit only a trend rather than a close approximation to a straight line. The extent to which the points are linearly related is specified quantitatively by the correlation coefficient. This is given by:

$$\rho = a*\sigma_x/\sigma_y \tag{A4.28}$$

where σ_x, σ_y are the variances of the x and y values about their mean. In the more general case where any function is used to describe y on x:

$$\rho = \sigma_{xy}/(\sigma_x\sigma_y) \tag{A4.29}$$

where σ_{xy} is the covariance of x and y given by:

$$\sigma_{xy} = \Sigma(x - m_x)*(y - m_y)/n \tag{A4.30}$$

and m_x, m_y are the means of the x and y values. The correlation coefficient will be 1 when the match is perfect between the model and the data, and zero if there is no correlation. In practice, it is usually easiest to do correlation within an Excel™ spreadsheet. A convenient way to do the fitting, where multiple variables and a nonlinear equation is being used is as follows:

Set trial values of the relevant coefficients in cells in the spreadsheet. Using these coefficients, calculate the model result (y') at all values of x for which a y value is available for comparison. In a new column calculate $(y - y')^2$ for each data point At the bottom of this column create the sum of all the values. The fit will obviously be optimized when the set of coefficients is found that minimizes this sum. This set can be found automatically within Excel™ using the Goal Seek™ function. Depending on the complexity of the equation and number of variables, it may be necessary to constrain the ranges of the coefficients. Excel™ can also return the correlation coefficient.

ABBREVIATIONS AND ACRONYMS

AHD	Alonghole (sometimes used in place of measured depth)
AI	Acoustic impedance
API	American Petroleum Institute (also units used for gamma ray)
AVO	Amplitude versus offset (of seismic traces)
B/D	Barrels per day
B_g	Gas volumetric factor (in scf under standard conditions per scf in reservoir)
BHA	Bottomhole [drilling] assembly
BHT	Bottomhole temperature
BOE	Barrel of oil equivalent
BOP	Blowout Preventer
bopd	Barrels of oil per day
BQv	The product of B, the equivalent counter-ion conductance, and Qv, the cation exchange capacity per unit pore volume.
BS	Bit size
BU	Buildup
BV	Bulk volume
BVI	Bulk volume of irreducible water
CAL	Caliper
CBL	Cement bond log
CCL	Casing collar locator
CEC	Cation exchange capacity
CHP	Casing head pressure
COI	Cost of information
CPI	Computer processed interpretation
DD	Driller's depth
DF	Derrick floor

DHI	Direct hydrocarbon indicator
DOP	Dropoff point
DT	Delta time (inverse of compressional velocity as measured by sonic tool)
DTS	Delta time shear
Ec	Eckert number
EHC	Equivalent hydraulic conductivity
EI	Elastic impedance
E_k	Kinetic energy
EOR	Enhanced oil recovery
ESP	Electric submersible pump
FBU	Formation buildup
FFI	Formation fluid index
FOL	Free oil level
FPI	Free point indicator
FRF	Formation resistivity factor
FSI	Formation strength indicator
FWL	Free water level
GBV	Gross bulk volume
GC	Gas chromatography
GDT	Gas down to
GIIP	Gas initially in place
GL	Ground level
GOC	Gas/oil contact
GOR	Gas/oil ratio
GR	Gamma ray
GUT	Gas up to
GWC	Gas/water contact
HCPV	Hydrocarbon pore volume
HDT	Higher dipmeter tool
HI	Hydrogen index
HIIP	Hydrocarbons initially in place
HUD	Holdup depth
HWC	Hydrocarbon/water contact
ID	Inner diameter
JV	Joint venture
k	Permeability
KB	Kelly bushing
KCl	Potassium chloride

k_h	Horizontal permeability
K_m	Matrix modulus
KOP	Kickoff point
k_v	Vertical permeability
LCM	Lost circulation material
LST	Limestone
LWD	Logging while drilling
MD	Measured depth
MWD	Measurement while drilling
NMR	Nuclear magnetic resonance
NRV	Net rock volume
NTG	Net to gross
OBM	Oil-based mud
OD	Outer diameter
ODT	Oil down to
OUT	Oil up to
OWC	Oil/water contact
P_c	Capillary pressure
Pe	Photoelectric effect (tool)
PE	Petroleum engineering
P_f	Formation pressure
PHIT	Matrix-corrected total porosity index
PI	Productivity index
psia	Pounds per square inch absolute
psig	Pounds per square inch per gauge
PT	Production technology
PU	Porosity units
PVT	Pressure, volume, temperature
Q	Flow rate
QA	Quality assurance
QC	Quality control
Q_v	Cation exchange capacity per unit pore volume
RB	Relative bearing
rb/stb	Reservoir barrels/stock tank barrels
rho_g	Grain density
R_m	Mud resistivity
R_{mf}	Mud filtrate resistivity
ROP	Rate of penetration

ROS	Residual oil saturation
RQI	Reservoir quality index
R_t	True resistivity
RU	Rig up
R_w	Water resistivity
SCAL	Special core analysis
SEM	Scanning electron miscroscopy
SG	Gas saturation
SH	Hydrocarbon saturation
SO	Oil saturation
SOR	Residual oil saturation
SP	Spontaneous potential
SPE	Society of Petroleum Engineers
SST	Sandstone
STOIIP	Stock tank oil initially in place
S_w	Water saturation
TC	Total carbon
TD	Total depth
TDA	Time domain analysis
Te	Echo spacing time for NMR logging
TG	Total gas
TOC	Top of cement
TVD	True vertical depth
TVDss	True vertical depth subsea
T_w	Polarization wait time for NMR logging
TWT	Two-way time
TZ	Time vs. depth
UTC	Ultimate technical cost
UV	Ultraviolet
VOI	Value of information
V_p	Compressional velocity
V_s	Shear velocity
V_{sh}	Shale volume
VSP	Vertical seismic profile
WC	Water cut
WL	Wireline
WOB	Weight on bit
WST	Well shoot test
WUT	Water up to

A P P E N D I X 6

USEFUL CONVERSION UNITS AND CONSTANTS

Depth
1 ft = 0.3048 m

Volume

1 barrel = 0.15899 cubic meters = 5.614 cubic feet = 42 US gallons

1 US gallon = 0.1337 cu ft

Pressure/Density

1 psi = 0.06895 bar = 0.068065 atm = 6895 N/m^2 (Pascals)

1 psi/ft = 2.3095 g/cc = 22.6 kPa/m
1 g/cc = 8.35 lb/gal

Temperature

Conversion from °C to °F: °C = (°F − 32)*5/9

°C = °K + 273.16
°Rankine = 1.8*°K = °F + 460

Permeability

1 darcy = 10^{-12} m^2

Table 6.1

Lithology	GR (API)	Rho$_b$ (g/cc)	CNL (NPU)	AC (µs/ft)	Resistivity (ohms)	Pe	ρ$_e$ g/cc	U (barn/cc)
Sandstone matrix	Low	2.65	−5 to −3	51–56	High	1.8	2.65	4.79
Limestone matrix	Low	2.71	0	44–48	High	5.1	2.71	13.77
Chlorite	High	2.79	30–40	50–150	Low	6.3	2.79	17.58
Illite	High	2.52	30–40	50–150	Low	1–5	2.52	8.69
Kaolinite	High	2.41	30–40	50–150	Low	1–5	2.41	4.41
Montmorillonite	High	2.12	30–40	50–150	Low	1–5	2.12	4.32
Dolomite	Low	2.88	2–6	39–44	High	3.1–3.2	2.86	9.00
Anhydrite	Low	2.98	0	45–55	High	5–5.1	2.96	14.95
Salt	Low	2.03	100	180–190	High	4.65		
Brine	0	1–1.1	100	189	0–∞	0.36	1.185	0.96
Oil	0	0.60–1.0	70–100	210–240	Very high	0.12–0.13	0.95–0.97	0.11–0.12
Gas	0	0.1–0.5	10–50	500–1500	Very high	Low	0.1–0.5	Low

Viscosity

1 centipoise (cp) = 10^{-3} Pa.s

Concentration

To convert ppm of [Cl$^-$] to [NaCl], you must multiply by:

molecular weight of NaCl/molecular weight of Cl
= (23 + 35)/35 = 1.65.

Other useful molecular weights are:

K (potassium): 39
Ca (calcium): 40
Mg (magnesium): 24
Br (bromium): 80

Note that 1 g/l = 1000 ppm

Conversion from Bottomhole to Surface Conditions

Oil: B_o = 1.2–1.6 reservoir barrels/stock tank barrels
Gas: B_g = 0.8–1.2 reservoir barrel/scf

Resistivity of Saline Solutions as a Function of Temperature

R_w is approximated by:

$$R_{w2} = R_{w1} * (T_1 + k)/(T_2 + k)$$

where k = 6.77 when T is in °F and 21.5 when T is in °C.

Properties of Some Common Lithologies

Note that $U = Pe*\rho_e$, where Pe is the photoelectric absorption index as displayed on a typical log and U is the volumetric photoelectric absortion index. U is typically used in forward modeling of multimineral models, since it is independent of porosity.

APPENDIX 7

CONTRACTOR TOOL MNEMONICS

Schlumberger Tool Mnemonics

Mnemonic	Type	Mode	Application	Description
CALAA	CCL	WIRELINE	Collar_Locator	Casing Anomaly Locator—AA
CALB	CCL	WIRELINE	Collar_Locator	Casing Anomaly Locator—B
CALC	CCL	WIRELINE	Collar_Locator	Casing Anomaly Locator—C
CALD	CCL	WIRELINE	Collar_Locator	Casing Anomaly Locator—D
CALF	CCL	WIRELINE	Collar_Locator	Casing Anomaly Locator—F
CALFA	CCL	WIRELINE	Collar_Locator	Casing Anomaly Locator—FA
CALGA	CCL	WIRELINE	Collar_Locator	Casing Anomaly Locator—GA
CALJ	CCL	WIRELINE	Collar_Locator	Casing Anomaly Locator—J
CALJB	CCL	WIRELINE	Collar_Locator	Casing Anomaly Locator—JB
CALM	CCL	WIRELINE	Collar_Locator	Casing Anomaly Locator—M
CALN	CCL	WIRELINE	Collar_Locator	Casing Anomaly Locator—N
CALQ	CCL	WIRELINE	Collar_Locator	Casing Anomaly Locator—Q
CALQA	CCL	WIRELINE	Collar_Locator	Casing Anomaly Locator—QA
CALQB	CCL	WIRELINE	Collar_Locator	Casing Anomaly Locator—QB
CALQC	CCL	WIRELINE	Collar_Locator	Casing Anomaly Locator—QC
CALQT	CCL	WIRELINE	Collar_Locator	Casing Anomaly Locator—QT
CALR	CCL	WIRELINE	Collar_Locator	Casing Anomaly Locator—R
CALS	CCL	WIRELINE	Collar_Locator	Casing Anomaly Locator—S
CALT	CCL	WIRELINE	Collar_Locator	Casing Anomaly Locator—T
CALU	CCL	WIRELINE	Collar_Locator	Casing Anomaly Locator—U
CALV	CCL	WIRELINE	Collar_Locator	Casing Anomaly Locator—V
CALW	CCL	WIRELINE	Collar_Locator	Casing Anomaly Locator—W
CALY	CCL	WIRELINE	Collar_Locator	Casing Anomaly Locator—Y
CALYA	CCL	WIRELINE	Collar_Locator	Casing Anomaly Locator—YA
CALZ	CCL	WIRELINE	Collar_Locator	Casing Anomaly Locator—Z
CBT	CBT	WIRELINE	Cement_Evaluation	Cement Bond Tool
CBTE	CBT	WIRELINE	Cement_Evaluation	Cement Bond Tool—E
CCL	CCL	WIRELINE	Collar_Locator	Casing Collar Locator
CCLAF	CCL	WIRELINE	Collar_Locator	Casing Collar Locator—AF
CCLAG	CCL	WIRELINE	Collar_Locator	Casing Collar Locator—AG
CCLAJ	CCL	WIRELINE	Collar_Locator	Casing Collar Locator—AJ

Mnemonic	Type	Mode	Application	Description
CCLAK	CCL	WIRELINE	Collar_Locator	Casing Collar Locator—AK
CCLAL	CCL	WIRELINE	Collar_Locator	Casing Collar Locator—AL
CCLAM	CCL	WIRELINE	Collar_Locator	Casing Collar Locator—AM
CCLAN	CCL	WIRELINE	Collar_Locator	Casing Collar Locator—AN
CCLAP	CCL	WIRELINE	Collar_Locator	Casing Collar Locator—AP
CCLAR	CCL	WIRELINE	Collar_Locator	Casing Collar Locator—AR
CCLL	CCL	WIRELINE	Collar_Locator	Casing Collar Locator
CCLLB	CCL	WIRELINE	Collar_Locator	Casing Collar Locator—LB
CCLN	CCL	WIRELINE	Collar_Locator	Casing Collar Locator
CCLNB	CCL	WIRELINE	Collar_Locator	Casing Collar Locator
CCLX	CCL	WIRELINE	Collar_Locator	Casing Collar Locator
CET	CET	WIRELINE	Cement_Evaluation	Cement Evaluation Tool
CETB	CET	WIRELINE	Cement_Evaluation	Cement Evaluation Tool
CETC	CET	WIRELINE	Cement_Evaluation	Cement Evaluation Tool
CETD	CET	WIRELINE	Cement_Evaluation	Cement Evaluation Tool
CETE	CET	WIRELINE	Cement_Evaluation	Cement Evaluation Tool
CETF	CET	WIRELINE	Cement_Evaluation	Cement Evaluation Tool
CETG	CET	WIRELINE	Cement_Evaluation	Cement Evaluation Tool
CETH	CET	WIRELINE	Cement_Evaluation	Cement Evaluation Tool
CETJ	CET	WIRELINE	Cement_Evaluation	Cement Evaluation Tool
CIT	CIT	WIRELINE	Casing_Inspection	Casing Inspection Tool
CITA	CIT	WIRELINE	Casing_Inspection	Casing Inspection Tool
CPET	CPET	WIRELINE	Casing_Inspection	Corrosion Protection Evaluation Tool
DCALA	CCL	WIRELINE	Collar_Locator	Digital Casing Collar Locator—A
ETT	ETT	WIRELINE	Casing_Inspection	Electromagnetic Thickness Tool
ETTD	ETT	WIRELINE	Casing_Inspection	Electromagnetic Thickness Tool
ETTDB	ETT	WIRELINE	Casing_Inspection	Electromagnetic Thickness Tool
FTGT	FTGT	WIRELINE	Casing_Inspection	Tubing Geometry Tool
FTGTB	FTGT	WIRELINE	Casing_Inspection	Tubing Geometry Tool—B
GFAB	GFA	WIRELINE	Nuclear	Surface Powered Gamma Ray (W5)
GFAC	GFA	WIRELINE	Nuclear	Surface Powered Gamma Ray (W7)
GPT	GPT	WIRELINE	Nuclear	Gamma Ray Perforating Tool
GPTA	GPT	WIRELINE	Nuclear	Gamma Ray Perforating Tool—A
GUN	GUN	WIRELINE	Perforating	GUN
GUN1	GUN	WIRELINE	Perforating	Perforating Gun
GUN2	GUN	WIRELINE	Perforating	Perforating Gun
GUN3	GUN	WIRELINE	Perforating	Perforating Gun
GUN4	GUN	WIRELINE	Perforating	Perforating Gun
GUN5	GUN	WIRELINE	Perforating	Perforating Gun
GUN6	GUN	WIRELINE	Perforating	Perforating Gun
GUN7	GUN	WIRELINE	Perforating	Perforating Gun
GUN8	GUN	WIRELINE	Perforating	Perforating Gun
GUN9	GUN	WIRELINE	Perforating	Perforating Gun
HCMT	HCMT	WIRELINE	Cement_Evaluation	HPHT Slim Cement Mapping Tool
HCMT-A	HCMT	WIRELINE	Cement_Evaluation	HPHT Slim Cement Mapping Tool (A)
MFCT	MFCT	WIRELINE	Casing_Inspection	Multi-Finger Caliper Tool
MPBC	MPBC	WIRELINE	Mechanical	Mechanical Plugback Cartridge
MPBCAA	MPBC	WIRELINE	Mechanical	Mechanical Plugback Cartridge—AA
MPBT	MPBT	WIRELINE	Mechanical	Mechanical Plug Back Tool
MPD	MPD	WIRELINE	Auxiliary	Magnetic Positioning Device

Mnemonic	Type	Mode	Application	Description
MPDH	MPD	WIRELINE	Auxiliary	Magnetic Positioning Device
MPSU	MPSU	WIRELINE	Mechanical	Mechanical Plugback Setting Unit
MPSUAA	MPSU	WIRELINE	Mechanical	Mechanical Plugback Setting Unit—AA
MPSUBA	MPSU	WIRELINE	Mechanical	Mechanical Plugback Setting Unit—BA
MPSUBB	MPSU	WIRELINE	Mechanical	Mechanical Plugback Setting Unit—BB
MPSUCA	MPSU	WIRELINE	Mechanical	Mechanical Plugback Setting Unit—CA
MPSUCB	MPSU	WIRELINE	Mechanical	Mechanical Plugback Setting Unit—CB
MWPS	MWPT	WIRELINE	Perforating	Measurement While Perforating Sonde
MWPT	MWPT	WIRELINE	Perforating	Measurement While Perforating Tool
NDT	NDT	WIRELINE	Nuclear	Neutron Depth Tool
NDTA	NDT	WIRELINE	Nuclear	Neutron Depth Tool—A
NDTB	NDT	WIRELINE	Nuclear	Neutron Depth Tool—B
PAT	PAT	WIRELINE	Casing_Inspection	Pipe Analysis Tool
PAT-G	PAT	WIRELINE	Casing_Inspection	Pipe Analysis Tool—GA
PATA	PAT	WIRELINE	Casing_Inspection	Pipe Analysis Tool—A
PATB	PAT	WIRELINE	Casing_Inspection	Pipe Analysis Tool—B
PATC	PAT	WIRELINE	Casing_Inspection	Pipe Analysis Tool C
PATD	PAT	WIRELINE	Casing_Inspection	Pipe Analysis Tool—D
PATE	PAT	WIRELINE	Casing_Inspection	Pipe Analysis Tool—E
PATG	PAT	WIRELINE	Casing_Inspection	Pipe Analysis Tool—G
PERFO	PERFO	WIRELINE	Perforating	Perforating (Dummy) Tool
PGGA	PGGT	WIRELINE	Nuclear	Powered Gun Gamma Ray—A
PGGB	PGGT	WIRELINE	Nuclear	Powered Gun Gamma Ray—B
PGGC	PGGT	WIRELINE	Nuclear	Powered Gun Gamma Ray—C
PGGCC	PGGT	WIRELINE	Nuclear	Powered Gun Gamma Ray—CC
PGGD	PGGT	WIRELINE	Nuclear	Powered Gun Gamma Ray—D
PGGT	PGGT	WIRELINE	Nuclear	Powered Gun Gamma Ray
PHAT	PHAT	WIRELINE	Casing_Inspection	Pit and Hole Analysis Tool
PLUG	PLUG	WIRELINE	Mechanical	PLUG
SAFE	SAFE	WIRELINE	Perforating	Slapper-Activated Firing Equipment
SCALA	CCL	WIRELINE	Collar_Locator	Casing Collar Locator
SCCL	SCCL	WIRELINE	Collar_Locator	Slim Casing Collar Locator
SCCL-A	SCCL	WIRELINE	Collar_Locator	Slim Casing Collar Locator—A
SCMT	SCMT	WIRELINE	Cement_Evaluation	Slim Cement Mapping Tool
SCMT-A	SCMT	WIRELINE	Cement_Evaluation	Slim Cement Mapping Tool—A
SPC	SPC	WIRELINE	Perforating	Selective Perforating Cartridge
SPCA	SPC	WIRELINE	Perforating	Selective Perforating Cartridge—A
SPCB	SPC	WIRELINE	Perforating	Selective Perforating Cartridge—B
SPGC	SPC	WIRELINE	Perforating	Selective Perforating Cartridge
SPPT	SPPT	WIRELINE	Mechanical	Production Packer Tool
SPPTA	SPPT	WIRELINE	Mechanical	Production Packer Tool
UCI	UCI	WIRELINE	Scanning	Ultrasonic Corrosion Imager
USIT	USIT	WIRELINE	Scanning	Ultrasonic Imaging Tool

Drilling/Back-off

Mnemonic	Type	Mode	Application	Description
BO	BO	WIRELINE	Special_Purpose	Back-off Tool
CERB	CERT	WIRELINE	Special_Purpose	Correlatable Electromagnetic Recovery Tool, 1–11/16 Inch
CERC	CERT	WIRELINE	Special_Purpose	Correlatable Electromagnetic Recovery Tool, 3–3/8 Inch

Mnemonic	Type	Mode	Application	Description
CERD	CERT	WIRELINE	Special_Purpose	Correlatable Electromagnetic Recovery Tool, 2–1/8 Inch
CERE	CERT	WIRELINE	Special_Purpose	Correlatable Electromagnetic Recovery Tool, 2–3/4 Inch
CERT	CERT	WIRELINE	Special_Purpose	Correlatable Electromagnetic Recovery Tool
CERTA	CERT	WIRELINE	Special_Purpose	Correlatable Electromagnetic Recovery Tool—A
CERTB	CERT	WIRELINE	Special_Purpose	Correlatable Electromagnetic Recovery Tool—B
CERTC	CERT	WIRELINE	Special_Purpose	Correlatable Electromagnetic Recovery Tool—C
FPIT	FPIT	WIRELINE	Special_Purpose	Free Point Indicator Tool
FPITA	FPIT	WIRELINE	Special_Purpose	Free Point Indicator Tool—A
FPITC	FPIT	WIRELINE	Special_Purpose	Free Point Indicator Tool—C (monocable)
FPITD	FPIT	WIRELINE	Special_Purpose	Free Point Indicator Tool—D
FPITE	FPIT	WIRELINE	Special_Purpose	Free Point Indicator Tool—E (monocable)
GSTA	GSTA	MWD	Combination	GeoSteering Tool
GSTA-CBB	GSTA	MWD	Combination	GeoSteering Tool—CBB
IAB	IAB	MWD	Geometry	Inclination at Bit Tool
IAB4-AA	IAB	MWD	Geometry	Inclination at Bit Tool
IAB4I-AA	IAB	MWD	Geometry	4.75″ Inclination at Bit Tool—AA
IAB4I-AB	IAB	MWD	Geometry	4.75″ Inclination at Bit Tool—AB
IAB6I-AA	IAB	MWD	Geometry	6.75″ Inclination at Bit Tool—AA
IAB8I-AA	IAB	MWD	Geometry	8.0″ Inclination at Bit Tool—AA
IAB9I-AA	IAB	MWD	Geometry	9.625″ Inclination at Bit Tool—AA
IDEAL_SURF	IDEAL_SURF	MWD	Auxiliary	Pseudo-tool for IDEAL Surface Data Acquisition
M10	MWD	MWD	MWD	M10 Navigational Sub (PowerPulse)
MWD	MWD	MWD	MWD	Measurement While Drilling
SHARP	SHARP	MWD	MWD	Slim Hole Retrievable MWD Tool
SLIM1	SLIM1	MWD	MWD	Slim Hole Retrievable MWD Tool
VBHA-AA	VIPER	MWD	MWD	VIPER Slimhole Coiled Tubing MWD Tool—AA
VIPER	VIPER	MWD	MWD	VIPER Slimhole Coiled Tubing MWD Tool

General

Mnemonic	Type	Mode	Application	Description
AMS	AMS	WIRELINE	Auxiliary	Auxiliary Measurement Sonde
AST	AST	WIRELINE	Scanning	Acoustic Scanner Tool
BGIC	BGIC	WIRELINE	Geometry	Borehole Geometry Interface Cartridge
BGS	BGS	WIRELINE	Geometry	Borehole Geometry Sonde
BGT	BGT	WIRELINE	Geometry	Borehole Geometry Tool
BGTC	BGT	WIRELINE	Geometry	Borehole Geometry Tool
BGTX	BGT	WIRELINE	Geometry	Borehole Geometry Tool
BHTV	BHTV	WIRELINE	Scanning	Borehole Televiewer
BTTA	BHTV	WIRELINE	Scanning	Borehole Televiewer—A
BTTB	BHTV	WIRELINE	Scanning	Borehole Televiewer—B

Mnemonic	Type	Mode	Application	Description
BTTC	BHTV	WIRELINE	Scanning	Borehole Televiewer—C
CALI	CALI	WIRELINE	Caliper	Generalized Caliper
CBTT	BHTV	WIRELINE	Scanning	Combinable Borehole Televiewer
CMT	CMT	WIRELINE	Scanning	Circumferential Microsonic Tool
ECD	ECD	WIRELINE	Caliper	Eccentered Caliper Device
ECDC	ECD	WIRELINE	Caliper	Eccentered Caliper Device—C
EDAC	EDAC	WIRELINE	Caliper	Eccentered Dual Axis Caliper
EMS	EMS	WIRELINE	Auxiliary	Environment Measurement Sonde
EMSA	EMS	WIRELINE	Auxiliary	Environmental Measurement Sonde—A
EMSB	EMS	WIRELINE	Auxiliary	Environmental Measurement Sonde—B
GCAD	GCAD	WIRELINE	Geometry	Guidance Continuous Tool Anchor Device
GCADA	GCAD	WIRELINE	Geometry	GCT Anchoring Device—A
GCADB	GCAD	WIRELINE	Geometry	GCT Anchoring Device—B
GCADC	GCAD	WIRELINE	Geometry	GCT Anchoring Device—C
GCT	GCT	WIRELINE	Geometry	Guidance Continuous Tool
GCTA	GCT	WIRELINE	Geometry	Guidance Continuous Tool—A
GCTAB	GCT	WIRELINE	Geometry	Guidance Continuous Tool—AB
GCTB	GCT	WIRELINE	Geometry	Guidance Continuous Tool—B
GCTBB	GCT	WIRELINE	Geometry	Guidance Continuous Tool—BB
GPIT	GPIT	WIRELINE	Geometry	General Inclinometry Tool
GPITB	GPIT	WIRELINE	Geometry	General Purpose Inclinometry Tool
HRCC	HRCC	WIRELINE	Caliper	HILT High Resolution Common Cartridge
MCD	MCD	WIRELINE	Caliper	Mechanical Caliper Device
MCDB	MCD	WIRELINE	Caliper	Mechanical Caliper Device—B
MCDD	MCD	WIRELINE	Caliper	Mechanical Caliper Device—D
MCDF	MCD	WIRELINE	Caliper	Mechanical Caliper Device—F
MCDG	MCD	WIRELINE	Caliper	Mechanical Caliper Device—G
NOSE	NOSE	WIRELINE	Auxiliary	Nose Orienting Scanning Equipment
NOSEA	NOSE	WIRELINE	Auxiliary	Nose Orienting Scanning Equipment—A
SBTTA	BHTV	WIRELINE	Scanning	Slim Hole Borehole Televiewer—A
SPCS	SPCS	WIRELINE	Caliper	Slim Powered Caliper Sonde
TCS	TCS	WIRELINE	Caliper	Through Tubing Caliper Sonde
TCSC	TCS	WIRELINE	Caliper	Through Tubing Caliper Sonde—C
TCSE	TCS	WIRELINE	Caliper	Through Tubing Caliper Sonde—E
TCSX	TCS	WIRELINE	Caliper	Through Tubing Caliper Sonde—X
UBI	UBI	WIRELINE	Scanning	Ultrasonic Borehole Imager
VCD	VCD	WIRELINE	Caliper	Caliper Device
VCDD	VCD	WIRELINE	Caliper	Caliper Device

Geology

Mnemonic	Type	Mode	Application	Description
FBST	FBST	WIRELINE	Scanning	Full Bore Scanner Tool
FBSTA	FBST	WIRELINE	Scanning	Full-Bore Scanner—A
FBSTB	FBST	WIRELINE	Scanning	Full-Bore Scanner—B
GHMA	GHMT	WIRELINE	Geomagnetism	Geological High Sensitivity Magnetic Tool A
GHMT	GHMT	WIRELINE	Geomagnetism	Geological High Sensitivity Magnetic Tool
HDT	HDT	WIRELINE	Dipmeter	High Resolution Dipmeter Tool

Mnemonic	Type	Mode	Application	Description
HDTD	HDT	WIRELINE	Dipmeter	High Resolution Dipmeter Tool—D
HDTE	HDT	WIRELINE	Dipmeter	High Resolution Dipmeter Tool—E
HDTF	HDT	WIRELINE	Dipmeter	High Resolution Dipmeter Tool—F
HDTG	HDT	WIRELINE	Dipmeter	High Resolution Dipmeter Tool—G
HDTJ	HDT	WIRELINE	Dipmeter	High Resolution Dipmeter Tool—J
MESC	MEST	WIRELINE	Scanning	Micro-Electrical Scanner Cartridge
MEST	MEST	WIRELINE	Scanning	Micro-Electrical Scanner Tool
MESTA	MEST	WIRELINE	Scanning	Micro-Electrical Scanner Tool—A
MESTB	MEST	WIRELINE	Scanning	Micro-Electrical Scanner Tool—B
MESTC	MEST	WIRELINE	Scanning	Micro-Electrical Scanner Tool—C
OBDT	OBDT	WIRELINE	Dipmeter	Oil Base Mud Dipmeter Tool
OBDTA	OBDT	WIRELINE	Dipmeter	Oil Base Mud Dipmeter Tool—A
OBDTAB	OBDT	WIRELINE	Dipmeter	Oil Base Mud Dipmeter Tool—AB
OBDTB	OBDT	WIRELINE	Dipmeter	Oil Base Mud Dipmeter Tool—B
OBMT	OBMT	WIRELINE	Scanning	Oil Base Mud Formation Imager Tool
OBMT-AA	OBMT	WIRELINE	Scanning	Oil Base Mud Formation Imager Tool—AA
OBMT-AB	OBMT	WIRELINE	Scanning	Oil Base Mud Formation Imager Tool—AB
OBMT-ABA	OBMT	WIRELINE	Scanning	Oil Base Mud Formation Imager Tool—ABA
OBMT-ABB	OBMT	WIRELINE	Scanning	Oil Base Mud Formation Imager Tool—ABB
OBMT-B	OBMT	WIRELINE	Scanning	Oil Base Mud Formation Imager Tool—B
SHDT	SHDT	WIRELINE	Dipmeter	Stratigraphic High Resolution Dipmeter Tool
SHDTA	SHDT	WIRELINE	Dipmeter	Stratigraphic High Resolution Dipmeter Tool
SHDTB	SHDT	WIRELINE	Dipmeter	Stratigraphic High Resolution Dipmeter Tool

Geophysics

Mnemonic	Type	Mode	Application	Description
CSAT	CSAT	WIRELINE	Acoustic	Seismic Acquisition Tool
CSAT1	CSAT	WIRELINE	Acoustic	Combined Seismic Acquisition Tool 1
CSAT2	CSAT	WIRELINE	Acoustic	Combined Seismic Acquisition Tool 2
CSAT3	CSAT	WIRELINE	Acoustic	Combined Seismic Acquisition Tool 3
CSAT4	CSAT	WIRELINE	Acoustic	Combined Seismic Acquisition Tool 4
CWRT	CWRT	WIRELINE	Acoustic	Cross Well Receiver Tool, used to receive cross-well seismic signals
CWRTA	CWRT	WIRELINE	Acoustic	Cross Well Receiver Tool—A
DSA	DSA	WIRELINE	Acoustic	Downhole Seismic Array
DSAA	DSA	WIRELINE	Acoustic	Downhole Seismic Array—A
DSAB	DSA	WIRELINE	Acoustic	Downhole Seismic Array—B
QSST	QSST	WIRELINE	Acoustic	Quick Shot Seismic Tool
QSSTB	QSST	WIRELINE	Acoustic	Quick Shot Seismic Tool—B
SAT	SAT	WIRELINE	Acoustic	Seismic Acquisition Tool
SATA	SAT	WIRELINE	Acoustic	Seismic Acquisition Tool—A
SATB	SAT	WIRELINE	Acoustic	Seismic Acquisition Tool—B
VSIT	VSIT	WIRELINE	Acoustic	Versatile Seismic Imager

Mnemonic	Type	Mode	Application	Description
VSIT-A	VSIT	WIRELINE	Acoustic	Versatile Seismic Imager
WSAM	WSAM	WIRELINE	Acoustic	Well Seismic Acquisition Surface Module
WST	WST	WIRELINE	Acoustic	Well Seismic Tool
WSTA	WST	WIRELINE	Acoustic	Seismic Acquisition Tool

Petrophysics

Mnemonic	Type	Mode	Application	Description
AACT	AACT	WIRELINE	Nuclear	Aluminium Activation Clay Tool
ACTC	AACT	WIRELINE	Nuclear	Aluminium Activation Clay Tool—C
ACTD	AACT	WIRELINE	Nuclear	Aluminium Activation Clay Tool—D
AND	ADN	MWD	Nuclear	Azimuthal Density Neutron Tool
ADN4AA	ADN	MWD	Nuclear	4.75 Inch Azimuthal Density Neutron Tool
ADN675	ADN	MWD	Nuclear	Azimuthal Density Neutron Tool, 6.75 inch
ADN6AA	ADN	MWD	Nuclear	6.75 Inch Azimuthal Density Neutron Tool
ADN6BA	ADN	MWD	Nuclear	6.75 Inch Azimuthal Density Neutron Tool
AGS	AGS	WIRELINE	Nuclear	Aluminum Gamma Ray Spectroscopy Sonde (same hardware as HNGS)
AGS-AA	AGS	WIRELINE	Nuclear	Aluminum Gamma Ray Spectroscopy Sonde—AA (same hardware as HNGS-AA)
AGS-BA	AGS	WIRELINE	Nuclear	Aluminum Gamma Ray Spectroscopy Sonde—BA (same hardware as HNGS-BA)
AGS_AA	AGS	WIRELINE	Nuclear	Aluminum Gamma Ray Spectroscopy Sonde—AA (same hardware as HNGS-AA)
AGS_BA	AGS	WIRELINE	Nuclear	Aluminum Gamma Ray Spectroscopy Sonde—BA (same hardware as HNGS-BA)
AIT	AIT	WIRELINE	Resistivity	Array Induction Imager Tool
AITB	AIT	WIRELINE	Resistivity	Array Induction Tool—B
AITC	AIT	WIRELINE	Resistivity	Array Induction Tool—C
AITH	AIT	WIRELINE	Resistivity	Array Induction Tool—H
AITS	AIT	WIRELINE	Resistivity	Array Induction Tool Slimhole
ALAT	ALAT	WIRELINE	Resistivity	Azimuthal Laterolog
ALATA	ALAT	WIRELINE	Resistivity	Azimuthal Laterolog—A
ALATB	ALAT	WIRELINE	Resistivity	Azimuthal Laterolog—B
APS	APS	WIRELINE	Nuclear	Accelerator Porosity Sonde
APS-AA	APS	WIRELINE	Nuclear	Accelerator Porosity Sonde—AA
APS-BA	APS	WIRELINE	Nuclear	Accelerator Porosity Sonde—BA (both CTS and DTS Telemetry)
APS-C	APS	WIRELINE	Nuclear	Accelerator Porosity Sonde—C
ARC	ARC	MWD	Resistivity	Array Compensated Resistivity— Gamma Ray Tool
ARC5AA	ARC	MWD	Resistivity	4.75 Inch Array Resistivity Compensated Tool—AA

Mnemonic	Type	Mode	Application	Description
ARC5AB	ARC	MWD	Resistivity	4.75 Inch Array Resistivity Compensated Tool—AB
ARC5BA	ARC	MWD	Resistivity	4.75 Inch Array Resistivity Compensated Tool—BA
BSP	SP	WIRELINE	Spontaneous	Potential Bridle
CDM6AA	CDR	MWD	Resistivity	6.75 Inch Compensated Dual Resistivity Tool
CDM6AB	CDR	MWD	Resistivity	6.75 Inch Compensated Dual Resistivity Tool
CDM8AA	CDR	MWD	Resistivity	8.25 Inch Compensated Dual Resistivity Tool
CDM8AB	CDR	MWD	Resistivity	8.25 Inch Compensated Dual Resistivity Tool
CDN	CDN	MWD	Nuclear	Compensated Density Neutron Tool
CDN650	CDN	MWD	Nuclear	Compensated Density Neutron Tool, 6.5 inch
CDN800	CDN	MWD	Nuclear	Compensated Density Neutron Tool, 8.0 inch
CDR	CDR	MWD	Resistivity	Compensated Dual Resistivity—Gamma Ray Tool
CDR475	CDR	MWD	Resistivity	Compensated Dual Resistivity—Gamma Ray Tool, 4.75 inch
CDR650	CDR	MWD	Resistivity	Compensated Dual Resistivity—Gamma Ray Tool, 6.5 inch
CDR675	CDR	MWD	Resistivity	Compensated Dual Resistivity—Gamma Ray Tool, 6.75 inch
CDR800	CDR	MWD	Resistivity	Compensated Dual Resistivity—Gamma Ray Tool, 8.0 inch
CDR825	CDR	MWD	Resistivity	Compensated Dual Resistivity—Gamma Ray Tool, 8.25 inch
CDR950	CDR	MWD	Resistivity	Compensated Dual Resistivity—Gamma Ray Tool, 9.5 inch
CFRT	CFRT	WIRELINE	Resistivity	Cased-hole Formation Resistivity
CFRT-C	CFRT	WIRELINE	Resistivity	Cased-hole Formation Resistivity—C
CGRS	CGRS	WIRELINE	Nuclear	Compact Gamma Ray Sonde
CMR	CMR	WIRELINE	Nuclear_Magnetic	Combinable Magnetic Resonance Tool
CMR-A	CMR	WIRELINE	Nuclear_Magnetic	Combinable Magnetic Resonance Tool—A
CMR-B	CMR	WIRELINE	Nuclear_Magnetic	Combinable Magnetic Resonance Tool—B
CMRT	CMR	WIRELINE	Nuclear_Magnetic	Combinable Magnetic Resonance Tool
CNT	CNT	WIRELINE	Nuclear	Compensated Neutron Tool
CNTA	CNT	WIRELINE	Nuclear	Compensated Neutron Tool—A
CNTD	CNT	WIRELINE	Nuclear	Compensated Neutron Tool—D
CNTE	CNT	WIRELINE	Nuclear	Compensated Neutron Tool—E
CNTF	CNT	WIRELINE	Nuclear	Compensated Neutron Tool—F
CNTG	CNT	WIRELINE	Nuclear	Compensated Neutron Tool—G
CNTH	CNT	WIRELINE	Nuclear	Compensated Neutron Tool—H
CNTS	CNT	WIRELINE	Nuclear	Compensated Neutron Tool—S
CSA1	CST	WIRELINE	Sampling	Bottom Gun Modified CST-A

Mnemonic	Type	Mode	Application	Description
CSG2	CST	WIRELINE	Sampling	Core Sample Taker
CST	CST	WIRELINE	Sampling	Core Sample Taker
CSTA	CST	WIRELINE	Sampling	Core Sample Taker
CSTAA	CST	WIRELINE	Sampling	Core Sample Taker
CSTB	CST	WIRELINE	Sampling	Core Sample Taker
CSTBA	CST	WIRELINE	Sampling	Core Sample Taker—BA
CSTC	CST	WIRELINE	Sampling	Core Sample Taker
CSTDA	CST	WIRELINE	Sampling	Core Sample Taker—DA
CSTG	CST	WIRELINE	Sampling	Core Sample Taker
CSTG2	CST	WIRELINE	Sampling	Core Sample Taker
CSTJ	CST	WIRELINE	Sampling	Core Sample Taker—J
CSTU	CST	WIRELINE	Sampling	Core Sample Taker
CSTV	CST	WIRELINE	Sampling	Core Sample Taker
CSTW	CST	WIRELINE	Sampling	Core Sample Taker
CSTY	CST	WIRELINE	Sampling	Core Sample Taker
CSTZ	CST	WIRELINE	Sampling	Core Sample Taker
CSX1	CST	WIRELINE	Sampling	Bottom Gun Modified CST-G
CSX2	CST	WIRELINE	Sampling	Middle Gun Modified CST-G
CSX3	CST	WIRELINE	Sampling	Top Gun Modified CST-G
CSZ1	CST	WIRELINE	Sampling	Bottom Gun Modified CST-Z
CSZ2	CST	WIRELINE	Sampling	Middle Gun Modified CST-Z
CSZ3	CST	WIRELINE	Sampling	Upper Gun Modified CST-Z
DIT	DIT	WIRELINE	Resistivity	Dual Induction Tool
DITB	DIT	WIRELINE	Resistivity	Dual Induction Tool—B
DITD	DIT	WIRELINE	Resistivity	Dual Induction Tool—D
DITE	DIT	WIRELINE	Resistivity	Dual Induction Tool (Phasor)
DITX	DIT	WIRELINE	Resistivity	Dual Induction Tool—D (with BGIC)
DLT	DLT	WIRELINE	Resistivity	Dual Laterolog Tool
DLTB	DLT	WIRELINE	Resistivity	Dual Laterolog Tool—B
DLTC	DLT	WIRELINE	Resistivity	Dual Laterolog Tool—C
DLTD	DLT	WIRELINE	Resistivity	Dual Laterolog Tool—D
DLTE	DLT	WIRELINE	Resistivity	Dual Laterolog Tool—E
DPT	DPT	WIRELINE	Dielectric	Dielectric Propagation Tool
DPTA	DPT	WIRELINE	Dielectric	Dielectric Propagation Tool
DPTB	DPT	WIRELINE	Dielectric	Dielectric Propagation Tool
DSLC	DSLT	WIRELINE	Acoustic	Digitizing Sonic Logging Cartridge
DSLT	DSLT	WIRELINE	Acoustic	Digitizing Sonic Logging Tool
DSLT-BA	DSLT	WIRELINE	Acoustic	Digitizing Sonic Logging Tool—BA
DSLT-BB	DSLT	WIRELINE	Acoustic	Digitizing Sonic Logging Tool—BB
DSLT-BC	DSLT	WIRELINE	Acoustic	Digitizing Sonic Logging Tool—BC
DSLT-H	DSLT	WIRELINE	Acoustic	Digitizing Sonic Logging Tool—H
DSLTBA	DSLT	WIRELINE	Acoustic	Digitizing Sonic Logging Tool—BA
DSLTBB	DSLT	WIRELINE	Acoustic	Digitizing Sonic Logging Tool—BB
DSLTBC	DSLT	WIRELINE	Acoustic	Digitizing Sonic Logging Tool—BC
DSST	DSST	WIRELINE	Acoustic	Dipole Shear Sonic Imager Tool
DSST-C	DSST	WIRELINE	Acoustic	Dipole Shear Sonic Imager Tool—C
DSSTA	DSST	WIRELINE	Acoustic	Dipole Shear Sonic Imager Tool—A
DSSTB	DSST	WIRELINE	Acoustic	Dipole Shear Sonic Imager Tool—B
DSSTC	DSST	WIRELINE	Acoustic	Dipole Shear Sonic Imager Tool—C
DST	DST	WIRELINE	Resistivity	Dual Laterolog Tool with SRT

Mnemonic	Type	Mode	Application	Description
DSTB	DST	WIRELINE	Resistivity	Dual Laterolog Tool with SRT
DSTD	DST	WIRELINE	Resistivity	Dual Laterolog Tool with SRT
DSTE	DST	WIRELINE	Resistivity	Dual Laterolog Tool with SRT
DWST	DWST	WIRELINE	Acoustic	Digital Waveform Sonic Tool
ECC	ECC	WIRELINE	Nuclear	Elemental Capture Spectroscopy Cartridge (supports ECS)
ECC-A	ECC	WIRELINE	Nuclear	Elemental Capture Spectroscopy Cartridge—A
ECS	ECS	WIRELINE	Nuclear	Elemental Capture Spectroscopy Sonde
ECS-A	ECS	WIRELINE	Nuclear	Elemental Capture Spectroscopy Sonde—A
ECSB	ECSC	WIRELINE	Nuclear	Elemental Capture Spectroscopy Cartridge—BA, DTS(FTB) Telemetry
ECSC	ECSC	WIRELINE	Nuclear	Elemental Capture Spectroscopy Cartridge (same hardware as NPLC)
ECSC-AA	ECSC	WIRELINE	Nuclear	Elemental Capture Spectroscopy Cartridge—AA, CTS(DTB) Telemetry
ECSC-BA	ECSC	WIRELINE	Nuclear	Elemental Capture Spectroscopy Cartridge—BA, DTS(FTB) Telemetry
EPT	EPT	WIRELINE	Dielectric	Electromagnetic Propagation Tool
EPTD	EPT	WIRELINE	Dielectric	Electromagnetic Propagation Tool—D
EPTE	EPT	WIRELINE	Dielectric	Electromagnetic Propagation Tool—E
EPTG	EPT	WIRELINE	Dielectric	Electromagnetic Propagation Tool—G
ES	ES	WIRELINE	Resistivity	Electrical Survey Tool
FGT	FGT	WIRELINE	Nuclear	Formation Gamma Gamma Tool
FGTC	FGT	WIRELINE	Nuclear	Formation Gamma Gamma Tool—C
FGTCA	FGT	WIRELINE	Nuclear	Formation Gamma Gamma Tool—CA
GNT	GNT	WIRELINE	Nuclear	Gamma Neutron Tool
GNTK	GNT	WIRELINE	Nuclear	Gamma Neutron Tool
GNTN	GNT	WIRELINE	Nuclear	Gamma Neutron Tool
GRA	GRA	WIRELINE	Nuclear	Geochemical Reservoir Analyzer
GRT	GRT	WIRELINE	Nuclear	Gamma Ray Tool
GRTC	GRT	WIRELINE	Nuclear	High Temperature Gamma Ray Tool, 1–3/8 Inch
GST	GST	WIRELINE	Nuclear	Gamma Spectroscopy Tool
GST-A	GST	WIRELINE	Nuclear	Gamma Spectroscopy Tool
HALS	HALS	WIRELINE	Resistivity	HILT Azimuthal Laterolog Sonde
HALS-BHALS	HALS	WIRELINE	Resistivity	HILT Azimuthal Laterolog Sonde—B
HALSB	HALS	WIRELINE	Resistivity	HILT Azimuthal Laterolog Sonde—B
HAPS-BA	APS	WIRELINE	Nuclear	HPHT Accelerator Porosity Sonde—
HAPS-C	APS	WIRELINE	Nuclear	HPHT Accelerator
HGNS	HGNS	WIRELINE	Nuclear	HILT Gamma-Ray Neutron Sonde
HILT	HILT	WIRELINE	Combination	High Resolution Integrated Logging Tool
HILTB	HILT	WIRELINE	Combination	High Resolution Integrated Logging Tool—B
HILTC	HILT	WIRELINE	Combination	High Resolution Integrated Logging Tool—CTS Stand-alone
HILTD	HILT	WIRELINE	Combination	High Resolution Integrated Logging Tool—DTS Combinable

Mnemonic	Type	Mode	Application	Description
HIT	HIT	WIRELINE	Resistivity	Hostile Array Induction Tool
HIT-A	HIT	WIRELINE	Resistivity	Hostile Array Induction Tool—A
HITA	HIT	WIRELINE	Resistivity	Hostile Array Induction Tool—A
HLDS	HLDS	WIRELINE	Nuclear	Hostile Litho-Density Sonde
HLDT	HLDT	WIRELINE	Nuclear	Hostile Environment Litho Density Tool
HLDTA	HLDT	WIRELINE	Nuclear	Hostile Environment Litho Density Tool—A
HNCC	HNCC	WIRELINE	Nuclear	Hostile Environment Nuclear Cartridge
HNCC-A	HNCC	WIRELINE	Nuclear	Hostile Environment Nuclear Cartridge—A
HNGC	HNGC	WIRELINE	Nuclear	Hostile Natural Gamma Ray Spectrometry Cartridge
HNGC-A	HNGC	WIRELINE	Nuclear	Hostile Natural Gamma Ray Spectrometry Cartridge—
HNGS	HNGS	WIRELINE	Nuclear	Hostile Natural Gamma Ray Spectrometry Sonde
HNGS-AA	HNGS	WIRELINE	Nuclear	Hostile Natural Gamma Ray Sonde—AA
HNGS-BA	HNGS	WIRELINE	Nuclear	Hostile Natural Gamma Ray Sonde—BA
HNGT	HNGT	WIRELINE	Nuclear	Hostile Natural Gamma Ray Spectrometry Tool
HNPL-BA	NPLC	WIRELINE	Nuclear	HPHT Nuclear Porosity Lithology Cartridge
HRDD	HRDD	WIRELINE	Nuclear	HILT High Resolution Density Device
HRGD	HRGD	WIRELINE	Combination	HILT High Resolution Resistivity Gamma-Ray Density
HRLA	HRLA	WIRELINE	Resistivity	High Resolution Laterolog Array
HRLA-A	HRLA	WIRELINE	Resistivity	High Resolution Laterolog Array—A
HRLT	HRLT	WIRELINE	Resistivity	High Resolution Laterolog Array Tool
HRLT-B	HRLT	WIRELINE	Resistivity	High Resolution Laterolog Array Tool—B
HRLT-C	HRLT	WIRELINE	Resistivity	High Resolution Laterolog Array Tool—C
HSGT	HSGT	WIRELINE	Nuclear	Hostile Environment Gamma Ray Tool
HSGTA	HSGT	WIRELINE	Nuclear	Hostile Environment Gamma Ray Tool—A
HSLT	DSLT	WIRELINE	Acoustic	HPHT Sonic Logging Tool
HSLT-A	DSLT	WIRELINE	Acoustic	HPHT Digitizing Sonic Logging Tool
ILTA	IRT	WIRELINE	Resistivity	Induction Logging Tool
IMPA	IMPA	MWD	Resistivity	Compensated Array Resistivity Tool
IMPA-AA	IMPA	MWD	Resistivity	4.75 Inch Prototype Compensated Array Resistivity
IMPA-AB	IMPA	MWD	Resistivity	4.75 Inch Production Compensated Array Resistivity Tool without IAB
IMPA-BA	IMPA	MWD	Resistivity	4.75 Inch Production Compensated Array Resistivity Tool with IAB
IRT	IRT	WIRELINE	Resistivity	Induction Resistivity Tool
IRTF	IRT	WIRELINE	Resistivity	Induction Resistivity Tool
IRTJ	IRT	WIRELINE	Resistivity	Induction Resistivity Tool

Mnemonic	Type	Mode	Application	Description
IRTK	IRT	WIRELINE	Resistivity	Induction Resistivity Tool
IRTL	IRT	WIRELINE	Resistivity	Induction Resistivity Tool
IRTM	IRT	WIRELINE	Resistivity	Induction Resistivity Tool
IRTN	IRT	WIRELINE	Resistivity	Induction Resistivity Tool
IRTQ	IRT	WIRELINE	Resistivity	Induction Resistivity Tool
IRTR	IRT	WIRELINE	Resistivity	Induction Resistivity Tool
IRTX	IRT	WIRELINE	Resistivity	Induction Resistivity Tool—Q (with BGIC)
ISONIC	ISONIC	MWD	Acoustic	LWD Sonic Tool
LDS	LDS	WIRELINE	Nuclear	Litho Density Sonde (for IPLT)
LDSC	LDSC	WIRELINE	Nuclear	Litho Density Cartridge (supports LDS or HLDS)
LDSC-A	LDSC	WIRELINE	Nuclear	Litho Density Cartridge—A
LDT	LDT	WIRELINE	Nuclear	Litho Density Tool
LDTA	LDT	WIRELINE	Nuclear	Litho Density Tool—A
LDTC	LDT	WIRELINE	Nuclear	Litho Density Tool—C
LDTD	LDT	WIRELINE	Nuclear	Litho Density Tool—D
LL3	LL3	WIRELINE	Resistivity	Laterolog 3 Tool (predates digital era)
LL7	LL7	WIRELINE	Resistivity	Laterolog 7 Sonde (predates digital era)
MCFL	MCFL	WIRELINE	Resistivity	Micro-Cylindrically Focused Logging Device
MDLT	MDLT	WIRELINE	Resistivity	Medium Dual Laterolog Tool
MDLTA	MDLT	WIRELINE	Resistivity	Medium Dual Laterolog Tool—A
MDST	MDST	WIRELINE	Resistivity	Medium Dual Laterolog SFL Tool
MDSTA	MDST	WIRELINE	Resistivity	Medium Dual Laterolog SFL Tool—A
MIST	MIST	WIRELINE	Nuclear	Multiple Isotope Spectroscopy Tool
MISTA	MIST	WIRELINE	Nuclear	Multiple Isotope Spectroscopy Tool—A
MISTB	MIST	WIRELINE	Nuclear	Multiple Isotope Spectroscopy Tool—B
MLL	MLL	WIRELINE	Resistivity	Microlaterologlog Tool (predates digital era)
MLT	MLT	WIRELINE	Resistivity	Microlog Tool
MLTA	MLT	WIRELINE	Resistivity	Microlog Tool
MLTAA	MLT	WIRELINE	Resistivity	Microlog Tool—AA
MPT	MPT	WIRELINE	Resistivity	Microlog Proximity Tool
MPTD	MPT	WIRELINE	Resistivity	Microlog Proximity Tool—D
MRWD	MRWD	MWD	Nuclear_Magnetic	Magnetic Resonance While Drilling
MRWD6-AA	MRWD	MWD	Nuclear_Magnetic	Magnetic Resonance While Drilling—AA
MSCT	MSCT	WIRELINE	Sampling	Mechanical Sidewall Coring Tool
MSCTA	MSCT	WIRELINE	Sampling	Mechanical Sidewall Coring Tool—A
MSGT	MSGT	WIRELINE	Nuclear	Scintillation Gamma Ray Tool, 2–3/4 inch
MSGTA	MSGT	WIRELINE	Nuclear	Scintillation Gamma Ray Tool—A
NDS	CDN	MWD	Nuclear	Neutron Density Sonde
NDS6AA	CDN	MWD	Nuclear	6.5 Inch Low Flow Compensated Density Neutron Tool
NDS6AB	CDN	MWD	Nuclear	6.5 Inch Low Flow Compensated Density Neutron Tool
NDS6BA	CDN	MWD	Nuclear	6.5 Inch High Flow Compensated Density Neutron Tool

Mnemonic	Type	Mode	Application	Description
NDS8AA	CDN	MWD	Nuclear	8.0 Inch Compensated Density Neutron Tool
NDS8AB	CDN	MWD	Nuclear	8.0 Inch Compensated Density Neutron Tool
NGS	NGS	WIRELINE	Nuclear	Natural Gamma Ray Spectrometry Sonde
NGSA	NGS	WIRELINE	Nuclear	Natural Gamma Ray Spectrometry Sonde
NGSB	NGS	WIRELINE	Nuclear	Natural Gamma Ray Spectrometry Sonde
NGT	NGT	WIRELINE	Nuclear	Natural Gamma Ray Spectrometry Tool
NGTA	NGT	WIRELINE	Nuclear	Natural Gamma Ray Spectrometry Tool—A
NGTB	NGT	WIRELINE	Nuclear	Natural Gamma Ray Spectrometry Tool—B
NGTC	NGT	WIRELINE	Nuclear	Natural Gamma Ray Spectrometry Tool—C
NGTD	NGT	WIRELINE	Nuclear	Natural Gamma Ray Spectrometry Tool—D
NGTE	NGT	WIRELINE	Nuclear	Natural Gamma Ray Spectrometry Tool—E
NGTF	NGT	WIRELINE	Nuclear	Natural Gamma Ray Spectrometry Tool—F
NMT	NMT	WIRELINE	Nuclear_Magnetic	Nuclear Magnetism Tool
NMTC	NMT	WIRELINE	Nuclear_Magnetic	Nuclear Magnetism Tool—C
NMTCA	NMT	WIRELINE	Nuclear_Magnetic	Nuclear Magnetism Tool—CA
NMTCB	NMT	WIRELINE	Nuclear_Magnetic	Nuclear Magnetism Tool—CB
NPLC	NPLC	WIRELINE	Nuclear	Nuclear Porosity Lithology Cartridge
NPLC-AA	NPLC	WIRELINE	Nuclear	Nuclear Porosity Lithology Cartridge—AA (CTS Telemetry)
NPLC-BA	NPLC	WIRELINE	Nuclear	Nuclear Porosity Lithology Cartridge—BA (DTS Telemetry)
NPLC-BB	NPLC	WIRELINE	Nuclear	Nuclear Porosity Lithology Cartridge—BB (DTS Telemetry, w/o HNGS board)
NPLT	NPLT	WIRELINE	Nuclear	Nuclear Porosity Lithology Tool
PCD	PCD	WIRELINE	Resistivity	Powered Caliper Device
PCDA	PCD	WIRELINE	Resistivity	Powered Caliper Device—A
PCDB	PCD	WIRELINE	Resistivity	Powered Caliper Device—B
PGT	PGT	WIRELINE	Nuclear	Formation Density Tool
PGTE	PGT	WIRELINE	Nuclear	Compensated Density Tool
PGTF	PGT	WIRELINE	Nuclear	Compensated Density Tool
PGTG	PGT	WIRELINE	Nuclear	Compensated Density Tool
PGTH	PGT	WIRELINE	Nuclear	Compensated Density Tool
PGTK	PGT	WIRELINE	Nuclear	Compensated Density Tool
PGTL	PGT	WIRELINE	Nuclear	Compensated Density Tool
PGTM	PGT	WIRELINE	Nuclear	Compensated Density Tool
PNT	PNT	WIRELINE	Nuclear	Sidewall Neutron Tool
PNTA	PNT	WIRELINE	Nuclear	Sidewall Neutron Tool—A
PNTB	PNT	WIRELINE	Nuclear	Sidewall Neutron Tool—B

Mnemonic	Type	Mode	Application	Description
PNTC	PNT	WIRELINE	Nuclear	Sidewall Neutron Tool—C
QAIT	HIT	WIRELINE	Resistivity	Slim Hostile Array Induction Tool—A
QCNT-A	CNT	WIRELINE	Nuclear	SlimHot Compensated Neutron Tool—A
QLDTA	SLDT	WIRELINE	Nuclear	SlimHot Litho-Density Tool—A
RAB	RAB	MWD	Resistivity	Azimuthal Laterolog—Gamma Ray Tool
RAB675	RAB	MWD	Resistivity	Azimuthal Laterolog—Gamma Ray Tool, 6.75 inch
RAB6AA	RAB	MWD	Resistivity	6.75 Inch Resistivity At the Bit Tool
RAB6B	RAB	MWD	Resistivity	6.75 Inch Resistivity At the Bit Tool
RAB825	RAB	MWD	Resistivity	Azimuthal Laterolog—Gamma Ray Tool, 8.25 inch
RAB8A	RAB	MWD	Resistivity	8.25 Inch Resistivity At the Bit Tool
RGM8A	CDR	MWD	Resistivity	8.0 Inch Compensated Dual Resistivity Tool
RGM8A	CDR	MWD	Resistivity	8.0 Inch Compensated Dual Resistivity Tool
RGM8AC	CDR	MWD	Resistivity	8.0 Inch Compensated Dual Resistivity Tool
RGM9AA	CDR	MWD	Resistivity	9.5 Inch Compensated Dual Resistivity Tool
RGM9AB	CDR	MWD	Resistivity	9.5 Inch Compensated Dual Resistivity Tool
RGS	CDR	MWD	Resistivity	Compensated Dual Resistivity—Gamma Ray
RGS6AA	CDR	MWD	Resistivity	6.5 Inch Compensated Dual Resistivity Tool
RST	RST	WIRELINE	Nuclear	Reservoir Saturation Tool
RSTA	RST	WIRELINE	Nuclear	Reservoir Saturation Tool—A
RSTB	RST	WIRELINE	Nuclear	Reservoir Saturation Tool—B
SAIT	SAIT	WIRELINE	Resistivity	Slimhole Array Induction Tool
SAIT-AA	SAIT	WIRELINE	Resistivity	Slimhole Array Induction Tool
SDT	SDT	WIRELINE	Acoustic	Sonic Digital Tool
SDTA	SDT	WIRELINE	Acoustic	Sonic Digital Tool—A
SDTB	SDT	WIRELINE	Acoustic	Sonic Digital Tool—B
SDTC	SDT	WIRELINE	Acoustic	Sonic Digital Tool—C
SDTE	SDT	WIRELINE	Acoustic	Sonic Digital Tool—E
SGT	SGT	WIRELINE	Nuclear	Scintillation Gamma Ray Tool
SGTE	SGT	WIRELINE	Nuclear	Scintillation Gamma Ray Tool
SGTEA	SGT	WIRELINE	Nuclear	Scintillation Gamma Ray Tool—EA
SGTEE	SGT	WIRELINE	Nuclear	Scintillation Gamma Ray Tool—EE
SGTFL	SGT	WIRELINE	Nuclear	Scintillation Gamma-Ray
SGTG	SGT	WIRELINE	Nuclear	Scintillation Gamma Ray Tool—G
SGTK	SGT	WIRELINE	Nuclear	Scintillation Gamma Ray Tool—K
SGTL	SGT	WIRELINE	Nuclear	Scintillation Gamma Ray Tool—L
SGTN	SGT	WIRELINE	Nuclear	Scintillation Gamma Ray Tool—N
SGTR	SGT	WIRELINE	Nuclear	Scintillation Gamma Ray Tool—R
SLDT	SLDT	WIRELINE	Nuclear	Slimhole Litho-Density Tool
SLDTA	SLDT	WIRELINE	Nuclear	Slimhole Litho-Density Tool—A
SLDTB	SLDT	WIRELINE	Nuclear	Slimhole Litho-Density Tool (Hostile Environment)

Mnemonic	Type	Mode	Application	Description
SLT	SLT	WIRELINE	Acoustic	Sonic Logging Tool
SLTJ	SLT	WIRELINE	Acoustic	Sonic Logging Tool
SLTL	SLT	WIRELINE	Acoustic	Sonic Logging Tool
SLTM	SLT	WIRELINE	Acoustic	Sonic Logging Tool
SLTN	SLT	WIRELINE	Acoustic	Sonic Logging Tool
SLTQ	SLT	WIRELINE	Acoustic	Sonic Logging Tool
SLTS	SLT	WIRELINE	Acoustic	Sonic Logging Tool
SLTT	SLT	WIRELINE	Acoustic	Sonic Logging Tool
SMRT	SMRT	WIRELINE	Resistivity	Slim Micro Resistivity Tool
SMRTA	SMRT	WIRELINE	Resistivity	Slim Micro Resistivity Tool—A
SNPD	SNPD	WIRELINE	Nuclear	Sidewall Neutron Tool
SON675	ISONIC	MWD	Acoustic	Sonic Tool, 6.75 inch
SON825	ISONIC	MWD	Acoustic	Sonic Tool, 8.25 inch
SP	SP	WIRELINE	Potential	Spontaneous Potential
SPA	SPA	WIRELINE	Potential	Spontaneous_Potential SP Adapter
SPAA	SPA	WIRELINE	Potential	Spontaneous_Potential SP Adapter—A
SPE	SPE	WIRELINE	Potential	Spontaneous_Potential SP Extender
SPEA	SPE	WIRELINE	Potential	Spontaneous_Potential SP Extender—A
SPIN	SP	WIRELINE	Potential	Spontaneous_Potential Dummy Tool for SP
SRT	SRT	WIRELINE	Resistivity	Micro Spherically Focused Resistivity Tool
SRTB	SRT	WIRELINE	Resistivity	Micro Spherically Focused Resistivity Tool—B
SRTC	SRT	WIRELINE	Resistivity	Micro Spherically Focused Resistivity Tool—C
SRTD	SRT	WIRELINE	Resistivity	Micro Spherically Focused Resistivity Tool—D
SRTX	SRT	WIRELINE	Resistivity	Micro Spherically Focused Resistivity Tool—X
SSGT	SSGT	WIRELINE	Nuclear	Scintillation Gamma Ray Tool
SSGTA	SSGT	WIRELINE	Nuclear	Scintillation Gamma Ray Tool—A
SSLT	SSLT	WIRELINE	Acoustic	SLIM Sonic Logging Tool
SSLTA	SSLT	WIRELINE	Acoustic	SLIM Sonic Logging Tool—A
SSLTAA	SSLT	WIRELINE	Acoustic	SLIM Sonic Logging Tool—AA
STCB	CST	WIRELINE	Sampling	Core Sample Taker
SWD8AA	ISONIC MWD	LWD	Acoustic	8.25 Inch Engineering Prototype Sonic While Drilling Tool
SWD8BA	ISONIC MWD	LWD	Acoustic	8.25 Inch Experimental Monopole Sonic While Drilling Tool
SWD8CA	ISONIC MWD	LWD	Acoustic	8.25 Inch Pilot Series Sonic While Drilling Tool
SWT	SWT	WIRELINE	Nuclear	Water Saturation Tool
SWTC	SWTC	WIRELINE	Nuclear	Water Saturation Tool Cartridge
SWTCA	SWTC	WIRELINE	Nuclear	Water Saturation Tool Cartridge—A
SWTS	SWTS	WIRELINE	Nuclear	Water Saturation Tool Sonde
SWTSA	SWTS	WIRELINE	Nuclear	Water Saturation Tool Sonde—A
SWTSB	SWTS	WIRELINE	Nuclear	Water Saturation Tool Sonde—B
SWTX	SWTX	WIRELINE	Nuclear	Water Saturation Equipment
SWTXA	SWTX	WIRELINE	Nuclear	Water Saturation Equipment—A

Mnemonic	Type	Mode	Application	Description
TSGT	SGT	WIRELINE	Nuclear	Gamma Ray Tool Production
Production Logging				
ASMT	ASMT	WIRELINE	Acoustic	Acoustic Spectrum Measuring Tool
BSDT	BSDT	WIRELINE	Acoustic	Bottom Sand Detection Tool
CFM	CFM	WIRELINE	Flowmeter	CPLT Flowmeter (may be a model of FBS or CFS)
CFM1	CFM	WIRELINE	Flowmeter	CPLT Flowmeter 1 (may be a model of FBS or CFS)
CFM2	CFM	WIRELINE	Flowmeter	CPLT Flowmeter 2 (may be a model of FBS or CFS)
CFS	CFS	WIRELINE	Flowmeter	Continuous Flowmeter Sonde
CFSF	CFS	WIRELINE	Flowmeter	Continuous Flowmeter Sonde—F
CFSH	CFS	WIRELINE	Flowmeter	Continuous Flowmeter Sonde—H
CFSJ	CFS	WIRELINE	Flowmeter	Continuous Flowmeter Sonde—J
CFSK	CFS	WIRELINE	Flowmeter	Continuous Flowmeter Sonde—K
CFSN	CFS	WIRELINE	Flowmeter	Continuous Flowmeter Sonde—N
CFSP	CFS	WIRELINE	Flowmeter	Continuous Flowmeter Sonde—P
CFSQ	CFS	WIRELINE	Flowmeter	Continuous Flowmeter Sonde—Q
CFSR	CFS	WIRELINE	Flowmeter	Continuous Flowmeter Sonde—R
CFSX	CFS	WIRELINE	Flowmeter	Continuous Flowmeter Sonde—X
CHMS	CHMS	WIRELINE	Pressure	CPLT HP Gauge
CPLC	CPLC	WIRELINE	Combination	Compact Production Logging Cartridge
CPLS	CPLS	WIRELINE	Combination	Compact Production Logging Sonde
CPLT	CPLT	WIRELINE	Combination	CTS Production Logging Tool
CPLTA	CPLT	WIRELINE	Combination	CTS Production Logging Tool—A
CPLTB	CPLT	WIRELINE	Combination	CTS Production Logging Tool—B
CPLTC	CPLT	WIRELINE	Combination	CTS Production Logging Tool—C
CRG	CRG	WIRELINE	Pressure	Pressure Gauge Tool (Flopetrol-Johnston)
DEFT	DEFT	WIRELINE	Flowmeter	Digital Entry and Fluid Imager Tool
DEFTA	DEFT	WIRELINE	Flowmeter	Digital Entry and Fluid Imager Tool—A
DEFTAB	DEFT	WIRELINE	Flowmeter	Digital Entry and Fluid Imager Tool—AB
DEFTAB_2	DEFT	WIRELINE	Flowmeter	Digital Entry and Fluid Imager Tool—AB (second DEFT-AB in toolstring)
DEFTA_2	DEFT	WIRELINE	Flowmeter	Digital Entry and Fluid Imager Tool—A (second DEFT-A in toolstring)
DEFTB	DEFT	WIRELINE	Flowmeter	Digital Entry and Fluid Imager Tool—B
DEFTC	DEFT	WIRELINE	Flowmeter	Digital Entry and Fluid Imager Tool—C
DEFTC_2	DEFT	WIRELINE	Flowmeter	Digital Entry and Fluid Imager Tool—C (second DEFT-C in toolstring)
DFIC	DFIC	WIRELINE	Flowmeter	Dual Flowmeter Interface Cartridge
DFICA	DFIC	WIRELINE	Flowmeter	Dual Flowmeter Interface Cartridge—A
DFICB	DFIC	WIRELINE	Flowmeter	Dual Flowmeter Interface Cartridge—B
DGT	DGT	WIRELINE	Gravity	Differential Gravity Tool
DGT-AA	DGT	WIRELINE	Gravity	Differential Gravity Tool—AA
EFM	EFM	WIRELINE	Flowmeter	Electrical Flowmeter Tool (Flopetrol-Johnston)
EXP	EXP	WIRELINE	Combination	External Surface Pressure and Temperature Measurements

Mnemonic	Type	Mode	Application	Description
FBDS	CFS	WIRELINE	Flowmeter	Full Bore Directional Spinner Flowmeter Sonde
FBDSA	CFS	WIRELINE	Flowmeter	Full Bore Directional Spinner Flowmeter Sonde—A
FBSA	CFS	WIRELINE	Flowmeter	Full Bore Spinner Flowmeter Sonde—A
FBSB	CFS	WIRELINE	Flowmeter	Full Bore Spinner Flowmeter Sonde—B
FBSC	CFS	WIRELINE	Flowmeter	Full Bore Spinner Flowmeter Sonde—C
FBSD	CFS	WIRELINE	Flowmeter	Full Bore Spinner Flowmeter Sonde—D
FBSE	CFS	WIRELINE	Flowmeter	Full Bore Spinner Flowmeter Sonde—E
FBSX	CFS	WIRELINE	Flowmeter	Full Bore Spinner Flowmeter Sonde—X
FSMT	FSMT	WIRELINE	Nuclear	Formation Subsidence Monitoring Tool
FSMT-A	FSMT	WIRELINE	Nuclear	Formation Subsidence Monitoring Tool—A
FSMT-B	FSMT	WIRELINE	Nuclear	Formation Subsidence Monitoring Tool—B
GHOST	DEFT	WIRELINE	Flowmeter	Gas Holdup Optical Sensing Tool
GHOST2	DEFT	WIRELINE	Flowmeter	Gas Holdup Optical Sensing Tool—A (2nd GHOST-A in tool string')
GHOSTA	DEFT	WIRELINE	Flowmeter	Gas Holdup Optical Sensing Tool—A
GHOSTA_2	DEFT	WIRELINE	Flowmeter	Gas Holdup Optical Sensing Tool—A (2nd GHOST-A in tool string')
GMS	GMS	WIRELINE	Fluid_Density	Gradiomanometer Sonde
GMSC	GMS	WIRELINE	Fluid_Density	Gradiomanometer Sonde
GMSD	GMS	WIRELINE	Fluid_Density	Gradiomanometer Sonde
GPPT	GPPT	WIRELINE	Nuclear	Gamma Neutron Tool
HCFS	HCFS	WIRELINE	Flowmeter	High Temperature Continuous Flowmeter Sonde
HCFSA	HCFS	WIRELINE	Flowmeter	High Temperature Continuous Flowmeter Sonde—A
HCFT	HCFT	WIRELINE	Flowmeter	Flowmeter Sonde
HCFTA	HCFT	WIRELINE	Flowmeter	Flowmeter Sonde
HMS	HMS	WIRELINE	Pressure	Hewlett Packard Manometer Sonde
HMSA	HMS	WIRELINE	Pressure	Hewlett Packard Manometer Sonde
HMSB	HMS	WIRELINE	Pressure	Hewlett Packard Manometer Sonde
HPA	HPA	WIRELINE	Pressure	Hewlett Packard Adaptor
HPAA	HPA	WIRELINE	Pressure	Hewlett Packard Adaptor
HPAB	HPA	WIRELINE	Pressure	Hewlett Packard Adaptor
HPXA	HPA	WIRELINE	Pressure	Hewlett Packard Adaptor
HTT	HTT	WIRELINE	Temperature	High Resolution Thermometer Tool
HTTA	HTT	WIRELINE	Temperature	High Resolution Thermometer Tool—A
HTTB	HTT	WIRELINE	Temperature	High Resolution Thermometer Tool—B
HTTC	HTT	WIRELINE	Temperature	High Resolution Thermometer Tool—C
HTTCA	HTT	WIRELINE	Temperature	High Resolution Thermometer Tool—CA
HUM	HUM	WIRELINE	Fluid_Density	Hold Up Meter
HUMA	HUM	WIRELINE	Fluid_Density	Hold Up Meter—A
HUMB	HUM	WIRELINE	Fluid_Density	Hold Up Meter—B
ISDT	ISDT	WIRELINE	Acoustic	Inline Sand Detection Tool (3rd Party, e.g. Fluenta Technology)
LEE_FM	LEE_FM	WIRELINE	Flowmeter	Flowmeter manufactured by Lee Tools, operated by Schlumberger

Mnemonic	Type	Mode	Application	Description
LIFT	LIFT	WIRELINE	Flowmeter	Local Impedance Flowmeter Tool
LIFTB	LIFT	WIRELINE	Flowmeter	Local Impedance Flowmeter Tool
MSRT	MSRT	PRODUCTION	Combination	Multi-Sensor Recorder Tool, includes flowmeter, pressure, and temperature sensors
MSRTA	MSRT	PRODUCTION	Combination	Multi-Sensor Recorder Tool—A
MSRTB	MSRT	PRODUCTION	Combination	Multi-Sensor Recorder Tool—B
MSRTC	MSRT	PRODUCTION	Combination	Multi-Sensor Recorder Tool—C
MSRTD	MSRT	PRODUCTION	Combination	Multi-Sensor Recorder Tool—D
MTSA	MTS	WIRELINE	Combination	Manometer Thermometer Sonde—A
MTSC	MTS	WIRELINE	Combination	Manometer Thermometer Sonde—C
NFD	NFD	WIRELINE	Nuclear	Nuclear Fluid Density Tool
NFDA	NFD	WIRELINE	Nuclear	Nuclear Fluid Density Tool—A
NFDB	NFD	WIRELINE	Nuclear	Nuclear Fluid Density Tool—B
NFDC	NFD	WIRELINE	Nuclear	Nuclear Fluid Density Tool—C
PBFS	CFS	WIRELINE	Flowmeter	Petal Basket Flowmeter Sonde
PBFSA	CFS	WIRELINE	Flowmeter	Petal Basket Flowmeter Sonde—A
PBFSB	CFS	WIRELINE	Flowmeter	Petal Basket Flowmeter Sonde—B
PBFSC	CFS	WIRELINE	Flowmeter	Petal Basket Flowmeter Sonde—C
PBFT	CFS	WIRELINE	Flowmeter	Petal Basket Flowmeter Tool
PBMS	PBMS	WIRELINE	Combination	PSP Basic Module Sonde
PFCS	PFCS	WIRELINE	Flowmeter	PSP Flowmeter Dual Caliper Sonde
PGMC	PGMC	WIRELINE	Fluid_Density	PSP Gradiomanometer Carrier
PGMC-A	PGMC	WIRELINE	Fluid_Density	PSP Gradiomanometer Carrier
PGMS	PGMS	WIRELINE	Fluid_Density	PSP Gradiomanometer Sonde
PILSA	CFS	WIRELINE	Flowmeter	PSP Flowmeter Dual Caliper Sonde
PMIT	PMIT	WIRELINE	Casing_Inspection	Multifinger Imaging Tool
PMIT-A	PMIT	WIRELINE	Casing_Inspection	Multifinger Imaging Tool—A
PMIT-B	PMIT	WIRELINE	Casing_Inspection	Multifinger Imaging Tool—B
PPS	PPS	WIRELINE	Testing	Production Packer Sonde, 1–11/16 Inch
PPSA	PPS	WIRELINE	Testing	Production Packer Sonde—A
PPSB	PPS	WIRELINE	Testing	Production Packer Sonde—B
PPT	PPT	WIRELINE	Testing	Production Packer Tool
PSPT	PSPT	WIRELINE	Combination	Production Services Logging Platform
PST	PST	WIRELINE	Sampling	Production Fluid Sampling Tool
PSTA	PST	WIRELINE	Sampling	Production Fluid Sampler Tool
PSTT	PSTT	WIRELINE	Mechanical	Production Services Tractor Tool
PSTT-A	PSTT	WIRELINE	Mechanical	Production Services Tractor Tool—A
PTS	PTS	WIRELINE	Combination	Pressure Temperature Sonde
PTSA	PTS	WIRELINE	Combination	Pressure Temperature Sonde—A
PTSAB	PTS	WIRELINE	Combination	Pressure Temperature Sonde—AB
PTSB	PTS	WIRELINE	Combination	Pressure Temperature Sonde—B
PUCS	PUCS	WIRELINE	Pressure	PSP Unigage Carrier Sonde
PUCS-A1	PUCS	WIRELINE	Combination	PSP Unigauge Carrier Sonde—A (first PUCS-A tool of combination)
PUCS-A2	PUCS	WIRELINE	Combination	PSP Unigauge Carrier Sonde—A (second PUCS-A tool of combination)
PVS	PVS	WIRELINE	Special_Purpose.	Phase Velocity Sonde: marker fluid ejector for phase velocity determination

Mnemonic	Type	Mode	Application	Description
PVS-AA	PVS	WIRELINE	Special_Purpose	Phase Velocity Sonde—AA: 1 11/16 inch marker fluid ejector
RCT	RCT	WIRELINE	Flowmeter	Flowmeter Transmitter (Rotron type)
RCTA	RCT	WIRELINE	Flowmeter	Flowmeter Transmitter (Rotron type)—A
SCTT	SCTT	WIRELINE	Temperature	Sidewall Contact Temperature Tool
SPG	SPG	WIRELINE	Pressure	Strain Pressure Gauge
SPST	SPST	WIRELINE	Sampling	Production Fluid Sampler Tool
SPSTA	SPST	WIRELINE	Sampling	Production Fluid Sampler Tool—A
SPTS	SPTS	WIRELINE	Combination	Pressure Temperature Sonde
SPTSA	SPTS	WIRELINE	Combination	Pressure Temperature Sonde—A
SVFS	SVFS	WIRELINE	Flowmeter	Slim Hole Vortex Flowmeter Sonde
SVFSA	SVFS	WIRELINE	Flowmeter	Slim Hole Vortex Flowmeter Sonde—A
TDMB	TDT	WIRELINE	Nuclear	Thermal Decay Time Tool
TDT	TDT	WIRELINE	Nuclear	Thermal Decay Time Tool
TDTK	TDT	WIRELINE	Nuclear	Thermal Decay Time Tool
TDTM	TDT	WIRELINE	Nuclear	Thermal Decay Time Tool
TDTP	TDT	WIRELINE	Nuclear	Thermal Decay Time Tool
TEMP	TEMP	WIRELINE	Temperature	Temperature
TET	TET	WIRELINE	Special_Purpose	Tracer Ejector Tool
TETD	TET	WIRELINE	Special_Purpose	Tracer Ejector Tool
TETE	TET	WIRELINE	Special_Purpose	Tracer Ejector Tool
TMT	TMT	WIRELINE	Temperature	Temperature Manometer Tool
TPT	TPT	WIRELINE	Combination	Temperature and Pressure Tool (Flopetrol-Johnston)

Well Testing

Mnemonic	Type	Mode	Application	Description
CP_1	MDCP	WIRELINE	Sampling	Modular Dynamics Casing Driller Probe Module 1
CP_2	MDCP	WIRELINE	Sampling	Modular Dynamics Casing Driller Probe Module 2
CP_3	MDCP	WIRELINE	Sampling	Modular Dynamics Casing Driller Probe Module 3
DP_1	MRDP	WIRELINE	Sampling	Modular Formation Dynamics Tester Multi-Probe Module
DP_2	MRDP	WIRELINE	Sampling	Modular Formation Dynamics Tester Multi-Probe Module
DP_3	MRDP	WIRELINE	Sampling	Modular Formation Dynamics Tester Multi-Probe Module
DWCS	DWCS	PRODUCTION	Special_Purpose	Deep Water Control System
DWCS-A	DWCS	PRODUCTION	Special_Purpose	Deep Water Control System—A
FC_1	MRFC	WIRELINE	Sampling	Modular Formation Dynamics Tester Flow Control Module
FC_2	MRFC	WIRELINE	Sampling	Modular Formation Dynamics Tester Flow Control Module
FC_3	MRFC	WIRELINE	Sampling	Modular Formation Dynamics Tester Flow Control Module
GFA	GFA	WIRELINE	Nuclear	Formation Tester Gamma Ray Detector
GFAA	GFA	WIRELINE	Nuclear	Formation Tester Gamma Ray Detector
GFT	GFT	WIRELINE	Nuclear	Formation Tester Gamma Ray
GFTA	GFT	WIRELINE	Nuclear	Formation Tester Gamma Ray

Mnemonic	Type	Mode	Application	Description
HY1	MRTT	WIRELINE	Sampling	MRTT Hydraulic Module
HY2	MRTT	WIRELINE	Sampling	MRTT Hydraulic Module
HY3	MRTT	WIRELINE	Sampling	MRTT Hydraulic Module
HY_1	MRHY	WIRELINE	Sampling	Modular Formation Dynamics Tester Hydraulic Module
HY_2	MRHY	WIRELINE	Sampling	Modular Formation Dynamics Tester Hydraulic Module
HY_3	MRHY	WIRELINE	Sampling	Modular Formation Dynamics Tester Hydraulic Module
MDCP	MDCP	WIRELINE	Sampling	Modular Dynamics Casing Driller Probe Module
MDT	MDT	WIRELINE	Sampling	Modular Formation Dynamics Tester
MP1	MRTT	WIRELINE	Sampling	MRTT Multi-Probe Module
MP2	MRTT	WIRELINE	Sampling	MRTT Multi-Probe Module
MP3	MRTT	WIRELINE	Sampling	MRTT Multi-Probe Module
MRDP	MRDP	WIRELINE	Sampling	MDT Dual-Probe Module
MRFA	MRFA	WIRELINE	Sampling	Optical Fluid Analyzer
MRFC	MRFC	WIRELINE	Sampling	MDT Flow Control Module
MRHY	MRHY	WIRELINE	Sampling	Modular Formation Dynamics Tester Hydraulic Module
MRMS	MRMS	WIRELINE	Sampling	MDT Multisample Module
MRPA	MRPA	WIRELINE	Sampling	MDT Dual-Packer Module
MRPC	MRPC	WIRELINE	Sampling	MDT Power Cartridge
MRPO	MRPO	WIRELINE	Sampling	MDT Pump-Out Module
MRPOUD	MRPOUD	WIRELINE	Sampling	MDT Up/Down Pump-Out Module
MRPS	MRPS	WIRELINE	Sampling	MDT Single-Probe Module
MRSC	MRSC	WIRELINE	Sampling	MDT Sample Module
MRTT	MRTT	WIRELINE	Sampling	Modular Reservoir Test Tool (MDT)
MS_1	MRMS	WIRELINE	Sampling	Multi-Sample Module (MRMS) 1
MS_2	MRMS	WIRELINE	Sampling	Multi-Sample Module (MRMS) 2
MS_3	MRMS	WIRELINE	Sampling	Multi-Sample Module (MRMS) 3
MS_4	MRMS	WIRELINE	Sampling	Multi-Sample Module (MRMS) 4
MS_5	MRMS	WIRELINE	Sampling	Multi-Sample Module (MRMS) 5
MTS	MTS	WIRELINE	Combination	Manometer Thermometer Sonde
PA	MRPA	WIRELINE	Sampling	Modular Formation Dynamics Tester Packer
PC	MRPC	WIRELINE	Sampling	Modular Formation Dynamics Tester Power Cartridge
PO	MRPO	WIRELINE	Sampling	Modular Formation Dynamics Tester Pumpout
POUD	MRPOUD	WIRELINE	Sampling	Modular Formation Dynamics Tester Up/Down Pumpout
PP1	MRTT	WIRELINE	Sampling	MRTT Precision Pressure Module
PP2	MRTT	WIRELINE	Sampling	MRTT Precision Pressure Module
PP3	MRTT	WIRELINE	Sampling	MRTT Precision Pressure Module
PQG	PQG	PRODUCTION	Special_Purpose	Permanent Quartz Pressure Gauge
PQG1	PQG	PRODUCTION	Special_Purpose	Permanent Quartz Pressure Gauge 1
PQG2	PQG	PRODUCTION	Special_Purpose	Permanent Quartz Pressure Gauge 2
PS1	MRTT	WIRELINE	Sampling	MRTT Single-Probe Module

Mnemonic	Type	Mode	Application	Description
PS2	MRTT	WIRELINE	Sampling	MRTT Single-Probe Module
PS3	MRTT	WIRELINE	Sampling	MRTT Single-Probe Module
PS_1	MRPS	WIRELINE	Sampling	Modular Formation Dynamics Tester Single-Probe
PS_2	MRPS	WIRELINE	Sampling	Modular Formation Dynamics Tester Single-Probe
PS_3	MRPS	WIRELINE	Sampling	Modular Formation Dynamics Tester Single-Probe
RFT	RFT	WIRELINE	Sampling	Repeat Formation Tester
RFTA	RFT	WIRELINE	Sampling	Repeat Formation Tester
RFTAB	RFT	WIRELINE	Sampling	Repeat Formation Tester
RFTB	RFT	WIRELINE	Sampling	Repeat Formation Tester
RFTTN	RFT	WIRELINE	Sampling	Repeat Formation Tester
RPQS	RPQS	WIRELINE	Sampling	Repeat Formation Tester Quartz Pressure Sonde
RTBC	RFT	WIRELINE	Sampling	Repeat Formation Tester Cased Hole Version
RTBO	RFT	WIRELINE	Sampling	Repeat Formation Tester Open Hole Version
RTCU	RFT	WIRELINE	Sampling	Repeat Formation Tester Cased Hole Version
RTOU	RFT	WIRELINE	Sampling	Repeat Formation Tester Open Hole Version
SC1	MRTT	WIRELINE	Sampling	MRTT Sample Chamber
SC_1	MRSC	WIRELINE	Sampling	Modular Formation Dynamics Tester Sample Chamber
SPFT	SPFT	WIRELINE	Testing	Slim Packer Fluid Analyzer Tool
SPFTA	SPFT	WIRELINE	Testing	Slim Packer Fluid Analyzer Tool
SRFT	RFT	WIRELINE	Sampling	Slim Repeat Formation Tester
SRFTA	RFT	WIRELINE	Sampling	Slim Repeat Formation Tester—A
SRFTB	RFT	WIRELINE	Sampling	Slim Repeat Formation Tester—B
SRFT_A	RFT	WIRELINE	Sampling	Slim Repeat Formation Tester—A
SRFT_B	RFT	WIRELINE	Sampling	Slim Repeat Formation Tester—B
SRFT_C	RFT	WIRELINE	Sampling	Slim Repeat Formation Tester—C
SRFT_D	RFT	WIRELINE	Sampling	Slim Repeat Formation Tester—D
SRFT_E	RFT	WIRELINE	Sampling	Slim Repeat Formation Tester—E
SRFT_F	RFT	WIRELINE	Sampling	Slim Repeat Formation Tester—F
WTPS	WTPS	WIRELINE	Pressure	Well Test Pressure Sonde
WTPT	WTPT	WIRELINE	Pressure	Well Test Pressure Tool

Halliburton Tool Mnemonics

Mnemonic	Tool	Description	Type	Code
2AC	2ACBA	TWO ARM CALIPER—DDL (older software versions)	CALIPER	XL1K
2AC	2ACCA	TWO ARM CALIPER—DDL (older software versions)	CALIPER	XL1K
ACCZ	ACCZ	ACCELEROMETER Z	AUXILIARY	PLS2
BATS	BATS	BOREHOLE AUDIO TEMPERATURE TOOL	COMPLETION	CHLS
BATS	BATSAA	BOREHOLE AUDIO TEMPERATURE TOOL	COMPLETION	XL1K
BCS	BCSBB	BOREHOLE COMPENSATED SONIC	SONIC	XL1K
BCS	BCSBD	BOREHOLE COMPENSATED SONIC	SONIC	XL1K
BCS	BCSD	DITS B/H COMPENSATED SONIC	SONIC	PLS2
BCS	BCSFA	BOREHOLE COMPENSATED SONIC	SONIC	XL1K
BCS	BCSHA	BOREHOLE COMPENSATED SONIC	SONIC	XL1K
BCS	BCSJA	BOREHOLE COMPENSATED SONIC	SONIC	XL1K
BCSD	BCSD	BOREHOLE COMPENSATED SONIC—DITS	SONIC	PLS2
BCSD	BCSDAA	BOREHOLE COMPENSATED SONIC—DITS	SONIC	XL1K
BHC_GR	BCS	BHC GAMMA TOOL	SONIC	PLS2
BHCS	BCS	BOREHOLE COMPENSATED SONIC	SONIC	PLS2
BHCS_D	BCS	BHC SONIC WITH DIGITAL PICKS	SONIC	PLS2
BHV	BHV	BHV COMPUTATION PANEL—	DUMMY TOOL	XL1K
BIP	BIPAA	DDL SUPER STACK BOTTOM ISOLATION SUB	RESISTIVITY	XL1K
BPOLAR	COSMOS	BIPLOAR PULSE TOOLS (CosMos)	COMPLETION	CHLS
BRID	DLLT	CABLE ELECTRODE BRIDLE	AUXILIARY	PLS2
BRID_R	DLLT	RIDGID CABLE ELECTRODE BRIDLE	AUXILIARY	PLS2
BRIDG	DLL	DLL CABLE ELECTRODE BRIDLE	AUXILIARY	PLS2
C_GR	DDL GR	GO GAMMA RAY TOOL—CAST CARD	NUCLEAR	PLS2
CAL	CAL	PANEL—XL1 CALIPER PROCESSING	CALIPER	XL1K
CAL_BB	SLDCAL	SLD_BB CALIPER	CALIPER	PLS2
CAL4DC	GO	GO 4 ARM CALIPER	CALIPER	PLS2
CALDC2	GO	GO 2 ARM CALIPER	CALIPER	PLS2
CALDC4	GO	GO 4 ARM CALIPER	CALIPER	PLS2
CALI_1	SDLT-A	SDL CALIPER	CALIPER	PLS2
CALI_2	M320	MSFL / MEL CALIPER	CALIPER	PLS2
CALI_3	M202	D202 CALIPER	CALIPER	PLS2
CALI_4	HSDLM-B	HOSTILE SDL CALIPER	CALIPER	PLS2
CALI_5	HFDT-A	HFDT CALIPER	CALIPER	PLS2
CALI_6	M320	MICRO GUARD / DLLT CALIPER	CALIPER	PLS2
CALISW	M123	SWN CALIPER	CALIPER	PLS2
CALMSF	MSFL	DDL MSFL CALIPER	CALIPER	PLS2
CAST	CAST	ACOUSTIC SCANNING TOOL: DIO #6	IMAGING	PLS2
CAST	CASTAA	CIRCUMFERENCIAL ACOUSTIC SCANNING TOOL	IMAGING	XL1K
CAST	CASTBA	CIRCUMFERENCIAL ACOUSTIC SCANNING TOOL	IMAGING	XL1K
CAST	CASTXX	CIRCUMFERENCIAL ACOUSTIC SCANNING TOOL—DITS	IMAGING	XL1K
CASTD	CAST	DITS ACOUSTIC SCANNING TOOL	IMAGING	PLS2
CASTDP	CAST	DITS CAST_D PET MODE	IMAGING	PLS2

Mnemonic	Tool	Description	Type	Code
CBL	CBL	CEMENT BOND TOOL	CEMENT EVAL.	PLS2
CBL	CBLDC	CEMENT BOND TOOL	CEMENT EVAL.	XL1K
CBL	CBLEA	CEMENT BOND TOOL	CEMENT EVAL.	XL1K
CBL	CBLEB	CEMENT BOND TOOL	CEMENT EVAL.	XL1K
CBL	CBLFA	CEMENT BOND TOOL—(1 11/16 TOOL)	CEMENT EVAL.	XL1K
CBL	CBLFB	CEMENT BOND TOOL—(1 11/16 TOOL)	CEMENT EVAL.	XL1K
CBL	CBLHA	CEMENT BOND TOOL—(MODULAR)	CEMENT EVAL.	XL1K
CBL_D	FWAT-A	M305B DITS SHORT BOND MODE	CEMENT EVAL.	PLS2
CCAT	CCAT	COMPENSATED CEMENT ATTEN TOOL	CEMENT EVAL.	PLS2
CCATCL	M214	M/C CASING COLLAR LOCATOR	COMPLETION	PLS2
CCATGR	M507	MULTICHANNEL GAMMA RAY	NUCLEAR	PLS2
CCL	CCLPA	CASING COLLAR LOCATOR—(PLT)	COMPLETION	XL1K
CCL	CCLPB	CASING COLLAR LOCATOR—(PLT)	COMPLETION	XL1K
CCL	CCLQA	CASING COLLAR LOCATOR—(DIGITAL)	COMPLETION	XL1K
CCL	CCLRA	CASING COLLAR LOCATOR—(MODULAR)	COMPLETION	XL1K
CCL	CCLUN	CASING COLLAR LOCATOR— (UNIVERSAL)	COMPLETION	XL1K
CCL	CCLWA	CASING COLLAR LOCATOR—(DIGITAL)	COMPLETION	XL1K
CCL_1	M214	M/C CASING COLLAR LOCATOR	COMPLETION	PLS2
CCL_PL	PCU	CASING COLLAR LOCATOR	COMPLETION	PLS2
CCLDC	COSMOS	DC CASING COLLAR LOCATOR	COMPLETION	PLS2
CCLDC1	BELL	DC CASING COLLAR LOCATOR	COMPLETION	PLS2
CCLGP	GPL	GPL CASING COLLAR LOCATOR	COMPLETION	PLS2
CCLHG	HGNC-A	HGNC CASING COLLAR LOCATOR	COMPLETION	PLS2
CCLPET	UCCL	PET TOOL MAGNETIC CCL	COMPLETION	PLS2
CCLPIB	COSMOS	DC CASING COLLAR LOCATOR—PIB	COMPLETION	PLS2
CCLSG	SGNC-A	SGNC CASING COLLAR LOCATOR	COMPLETION	PLS2
CCLT	CCLTAA	CASING COLLAR LOCATOR—(DITS)	COMPLETION	XL1K
CDL	CDLGA	COMPENSATED DENSITY LOG TOOL— (3 3/8 TOOL)	NUCLEAR	XL1K
CDL	CDLKA	COMPENSATED DENSITY LOG TOOL	NUCLEAR	XL1K
CDL	CDLKB	COMPENSATED DENSITY LOG TOOL	NUCLEAR	XL1K
CDL	CDLLA	COMPENSATED DENSITY LOG TOOL— (2 3/4 TOOL)	NUCLEAR	XL1K
CDL	CDLMA	COMPENSATED DENSITY LOG TOOL	NUCLEAR	XL1K
CDL	CDLNA	COMPENSATED DENSITY LOG TOOL	NUCLEAR	XL1K
CDT	CDT	COMPENSATED DENSITY	NUCLEAR	PLS2
CDTCAL	CDTCAL	CDT CALIPER	CALIPER	PLS2
CH_G	CHTN	GO DOWNHOLE LOADCELL	AUXILIARY	PLS2
CH_HOS	CHTN	HOSTILE CABLE HEAD LOAD CELL	AUXILIARY	PLS2
CH_TEN	CHTN	DITS CABLE HEAD LOAD CELL	AUXILIARY	PLS2
CH2TEN	CHTN	D2TS CABLE HEAD LOAD CELL	AUXILIARY	PLS2
CHARM	CHARM	CASED HOLE ANALYSIS RESERVOIR MODEL	ANALYSIS	PLS2
CHFW	FWAT-A	M305 ACOUSTIC	SONIC	PLS2
CHS	FWAT-A	M305B CASED HOLE XTRA LS 4 RCVR DT	SONIC	PLS2
CHSF	CHSFAA	CASED HOLE—SEQ. FORMATION TESTER—DDL	PRODUCTION	XL1K
CIC	PENGO820	CASING INSPECTION CALIPER	CASING INSPECT.	PLS2

Mnemonic	Tool	Description	Type	Code
CIT	CIT	CASING INSPECTION TOOL	CASING INSPECT.PLS2	PLS2
CIT	CITAA	CASING INSPECTION TOOL	CASING INSPECT.	XL1K
CIT_A	CIT	CASING INSPECTION (ACOUS COMB)	CASING INSPECT.PLS2	PLS2
CLAMS	CLAMS	CLAY AND MATRIX ANALYSIS	ANALYSIS	PLS2
CNT	CNTAA	COMPENSATED NEUTRON—PAD STYLE	NUCLEAR	XL1K
CNT	CNTAB	COMPENSATED NEUTRON—PAD STYLE	NUCLEAR	XL1K
CNT	CNTBA	COMPENSATED NEUTRON—PAD STYLE	NUCLEAR	XL1K
CNT	CNTCA	COMPENSATED NEUTRON—PAD STYLE	NUCLEAR	XL1K
CNT	CNTDA	COMPENSATED NEUTRON—PAD STYLE	NUCLEAR	XL1K
CNT	CNTEA	COMPENSATED NEUTRON—PAD STYLE	NUCLEAR	XL1K
CNT	CNTFA	COMPENSATED NEUTRON—(2 3/4 TOOL)	NUCLEAR	XL1K
CNT	CNTJA	COMPENSATED NEUTRON—MANDREL	NUCLEAR	XL1K
CNT	CNTKA	COMPENSATED NEUTRON—MANDREL	NUCLEAR	XL1K
CNT	CNTKB	COMPENSATED NEUTRON—MANDREL	NUCLEAR	XL1K
CNT	CNTLA	COMPENSATED NEUTRON—MANDREL	NUCLEAR	XL1K
CNT	CNTMA	COMPENSATED NEUTRON—MANDREL	NUCLEAR	XL1K
CNT	CNTNA	COMPENSATED NEUTRON—MANDREL	NUCLEAR	XL1K
CNT	CNTPA	COMPENSATED NEUTRON—MANDREL	NUCLEAR	XL1K
CNT_K	CNTKA	COMPENSATED NEUTRON K MUX-E	NUCLEAR	PLS2
CNT_N	CNTNA	COMPENSATED NEUTRON MODEL N	NUCLEAR	PLS2
COM	COM	XL1—COMPUTED OUTPUTS PANEL	ANALYSIS	XL1K
CORAL	CORAL	COMPLEX RESERVOIR ANALYSIS MODEL	ANALYSIS	PLS2
CORE	SWC	M6 CORE GUN WITH SP	SAMPLING	PLS2
CP_CAL	COMPROBE	COMPROBE CALIPER	CALIPER	PLS2
CP_DEN	COMPROBE	COMPROBE DENSITY	NUCLEAR	PLS2
CP_DN	COMPROBE	COMPROBE DUAL NEUTRON	NUCLEAR	PLS2
CP_GR	COMPROBE	COMPROBE GAMMA/CALIPER/CCL	NUCLEAR	PLS2
CQPT	CQPTAA	COMPENSATED QUARTZ PRESSURE TOOL	PRODUCTION	XL1K
CSNG	CSNG	COMPENSATED SPECTRAL NATURAL GAMMA	NUCLEAR	CHLS
CSNG	CSNGGR	COMPENSATED SPECTRAL NATURAL GAMMA	NUCLEAR	XL1K
CSNG	CSNGMI	MINI TRACERSCAN—SLIM HOLE	NUCLEAR	XL1K
CSNG	CSNGTI	COMPENSATED SPECTRAL NATURAL GAMMA	NUCLEAR	XL1K
CSNG_G	CSNGG-A	DITS SPECTRAL GAMMA—GRAPHITE	NUCLEAR	PLS2
CSNG_T	CSNGT-A	DITS SPECTRAL GAMMA—TITANIUM	NUCLEAR	PLS2
CSNGMC	CSNG-MC	SPECTRAL GAMMA—GRAPHITE	NUCLEAR	PLS2
D2TS	D2TSAA	DITS 2 TELEMETRY SUB	TELEMETRY	XL1K
DC_CAL	DC CALIP	GO DC CALIPER	CALIPER	PLS2
DCHT	DCHTAA	DITS CABLEHEAD TENSION SUB	AUXILIARY	XL1K
DIEL	DIEL	DIELECTRIC TOOL	RESISTIVITY	PLS2
DIELGR	DIEL	DIELECTRIC TOOL GAMMA RAY	RESISTIVITY	PLS2
DIKA	DIKA	LATEROLOG 3 DIK-A	RESISTIVITY	PLS2
DIL	DIL	DUAL INDUCTION	RESISTIVITY	PLS2
DIL	DILAA	DUAL INDUCTION	RESISTIVITY	XL1K

Mnemonic	Tool	Description	Type	Code
DIL	DILBA	DUAL INDUCTION	RESISTIVITY	XL1K
DIL	DILBB	DUAL INDUCTION	RESISTIVITY	XL1K
DIL	DILBC	DUAL INDUCTION	RESISTIVITY	XL1K
DIL	DILBD	DUAL INDUCTION	RESISTIVITY	XL1K
DIL	DILCA	DUAL INDUCTION	RESISTIVITY	XL1K
DIL	DILCB	DUAL INDUCTION	RESISTIVITY	XL1K
DIL	DILCC	DUAL INDUCTION	RESISTIVITY	XL1K
DIL	DILDA	DUAL INDUCTION	RESISTIVITY	XL1K
DIL	DILEA	DUAL INDUCTION	RESISTIVITY	XL1K
DIL	DILFA	DUAL INDUCTION	RESISTIVITY	XL1K
DIL	DILGA	DUAL INDUCTION	RESISTIVITY	XL1K
DIL	DILHA	DUAL INDUCTION	RESISTIVITY	XL1K
DILT	DILTAA	DITS DUAL INDUCTION—(W/MGRD)	RESISTIVITY	XL1K
DIND	DILT-A B	DITS DUAL INDUCTION	RESISTIVITY	PLS2
DIOHD	CHTN	SVC HEADER/ SIG COND/ DIO STAT	DUMMY TOOL	PLS2
DIP_MC	M242-A	M/C DIPMETER INSTRUMENT	DIPMETER	PLS2
DIPCOR	DIPCOR	FOUR ARM DIPMETER ANALYSIS (for MC DIP tools)	ANALYSIS	PLS2
DITCCL	DCCL-A	DITS CCL	COMPLETION	PLS2
DITSHD	DUMMY	DITS SERVICE HEADER + ANALOG	DUMMY TOOL	PLS2
DLD	DLDAA	DDL DLL LOWER ELECTRODE	RESISTIVITY	XL1K
DLL	AA-EB	DDL STANDARD DUAL LATEROLOG	RESISTIVITY	PLS2
DLL	DLLAA	DDL STANDARD DUAL LATEROLOG	RESISTIVITY	XL1K
DLL	DLLBA	DDL STANDARD DUAL LATEROLOG	RESISTIVITY	XL1K
DLL	DLLBB	DDL STANDARD DUAL LATEROLOG	RESISTIVITY	XL1K
DLL	DLLCA	DDL STANDARD DUAL LATEROLOG	RESISTIVITY	XL1K
DLL	DLLDA	DDL STANDARD DUAL LATEROLOG	RESISTIVITY	XL1K
DLL	DLLEA	DDL STANDARD DUAL LATEROLOG	RESISTIVITY	XL1K
DLL	DLLEB	DDL STANDARD DUAL LATEROLOG	RESISTIVITY	XL1K
DLL	DLLFA	DDL EXTENDED DUAL LATEROLOG	RESISTIVITY	XL1K
DLL	DLLGA	DDL EXTENDED DUAL LATEROLOG	RESISTIVITY	XL1K
DLL	DLLHA	DDL EXTENDED DUAL LATEROLOG	RESISTIVITY	XL1K
DLLT	DLLT-A	DITS DUAL LATEROLOG	RESISTIVITY	PLS2
DLLT	DLLTAA	DITS DUAL LATEROLOG	RESISTIVITY	XL1K
DLLT	DLLTBA	DITS DUAL LATEROLOG	RESISTIVITY	XL1K
DLLX	FA,GA,HA	DDL EXTENDED DUAL LATEROLOG	RESISTIVITY	PLS2
DMSFL	DMSFL	STANDALONE DITS MSFL	RESISTIVITY	PLS2
DMSFLC	MSFL	DITS STANDALONE MSFL / MICLOG	RESISTIVITY	PLS2
DMSFLX	MSFL	DITS STANDALONE MSFL (EXP)	RESISTIVITY	PLS2
DSEN	DSEN-A	DUAL SPACED EPITHERMAL NEUTRON	NUCLEAR	PLS2
DSEN	DSENAA	DUAL SPACED EPITHERMAL NEUTRON	NUCLEAR	XL1K
DSN_II	DSNT-A	DUAL SPACED NEUTRON II	NUCLEAR	PLS2
DSNT	DSNTAA	DUAL SPACED NEUTRON II	NUCLEAR	XL1K
DSTU	DSTUAA	DITS SUBSURFACE TELEMETRY UNIT	TELEMETRY	XL1K
DSTU	DSTUBA	DITS SUBSURFACE TELEMETRY UNIT	TELEMETRY	XL1K
DTD	DTDAA	DOWNHOLE LINE TENSION	AUXILIARY	XL1K
DTD	DTDBA	DOWNHOLE LINE TENSION	AUXILIARY	XL1K
DTEN	CHTN	DIFFERENTIAL TENSION	AUXILIARY	PLS2
EMI2	EMI-B	DITS ELECTRIC MICRO IMAGER— (SINGLE DITS)	IMAGING	PLS2

Mnemonic	Tool	Description	Type	Code
EMI2CL	EMI-A	EMI CALIPER PACKAGE	CALIPER	PLS2
EMI2DP	EMI-A	EMI DIPMETER PACKAGE	RESISTIVITY	PLS2
EMI2MG	EMI-A	EMI NAVIGATION MAGNETOMETER	AUXILIARY	PLS2
EMID2	EMI-B	DITS ELECTRIC MICRO IMAGER— (DOUBLE DITS)	IMAGING	PLS2
EMIRES	EMI-B	EMI RESISTIVITY PACKAGE	RESISTIVITY	PLS2
EVR_DN	DSNT-A	DUAL SPACED NEUTRON II (EVR-II)	NUCLEAR	PLS2
EVR_GR	NGRT-A	GAMMA RAY TOOL (EVR-II)	NUCLEAR	PLS2
EVRSD8	SDLT-A	SPECTRAL DENSITY (8 BIT) (EVR-II)	NUCLEAR	PLS2
EVRSDL	SDLT-A	SPECTRAL DENSITY (EVR-II)	NUCLEAR	PLS2
FACMAN	FACT-A	FACT CALIPER MANDREL	CALIPER	PLS2
FACT	FACTAA	FOUR ARM CALIPER TOOL—DITS	CALIPER	XL1K
FDF	FDFAA	FLOW DIVERTER FLOWMETER	PRODUCTION	XL1K
FDT	FDTEA	FLUID DENSITY—MUXB PL	PRODUCTION	XL1K
FDT	FDTEB	FLUID DENSITY—MUXB PL	PRODUCTION	XL1K
FDT	FDTEC	FLUID DENSITY—MUXB PL	PRODUCTION	XL1K
FED	FED	DIO FOUR ELECTRODE DIPMETER	DIPMETER	PLS2
FED	FEDGA	FOUR ELECTRODE DIPMETER (4 1/2INCH)	DIPMETER	XL1K
FED	FEDHA	FOUR ELECTRODE DIPMETER (3 1/2 IN)	DIPMETER	XL1K
FED	FEDJA	FOUR ELECTRODE DIPMETER (4 1/2 IN)	DIPMETER	XL1K
FEDNAV	FED	FED NAVIGATION PACKAGE (G)	DIPMETER	PLS2
FHY	FHYGA	FULL BORE HYDRO TOOL	PRODUCTION	XL1K
FIAC	FIAC-A	FOUR INDEPENDENT ARM CALIPER	CALIPER	PLS2
FIAC	FIACAA	FOUR INDEPENDENT ARM CALIPER	CALIPER	XL1K
FLD_PL	PCU	FLUID DENSITY	PRODUCTION	PLS2
FLDN	BELL	DC FLUID DENSITY	PRODUCTION	PLS2
FLTT	M139-A	FLUID TRAVEL TOOL	PRODUCTION	PLS2
FMS	FMSHA	HI RES FLOWMETER SPINNER	PRODUCTION	XL1K
FMS	FMSHB	HI RES FLOWMETER SPINNER	PRODUCTION	XL1K
FMS	FMSHC	HI RES FLOWMETER SPINNER	PRODUCTION	XL1K
FWL2DT	FWAT-A	M305A 2 RCVR LONG SPACED—SNR-DT	SONIC	PLS2
FWS	FWAT-A	M305B 4 RCVR AUTO GAIN	SONIC	PLS2
FWS_D2	FWAT-A	M305B 4 RCVR AUTO GAIN—QUAD STACK	SONIC	PLS2
FWST	FWSTAA	M305A FULL WAVE SONIC	SONIC	XL1K
FWST2	FWAT-A	M305A ACOUSTIC	SONIC	PLS2
FWST23	FWAT-A	M305A THREE RECEIVER LONG SPACE	SONIC	PLS2
FWST2A	FWAT-A	M305A ACOUSITIC/AMPL	SONIC	PLS2
FWST2S	FWAT-A	M305A TWO RECEIVER LONG SPACED	SONIC	PLS2
FWST2U	FWAT-A	M305A 2 RCVR LONG SPACED—DT	SONIC	PLS2
FWST4	FWAT-A	M305A ACOUSTIC SHORT TIP MODE 4	SONIC	PLS2
FWST4A	FWAT-A	M305A ACOUSTIC SHORT BOND MODE	SONIC	PLS2
FWSTA	FWAT-A	M305A ACOUSTIC	SONIC	PLS2
FWSTA8	FWAT-A	M305A ACOUSTIC	SONIC	PLS2
FWSTAU	FWAT-A	M305A ACOUSTIC (MODE A) FOR INT'L	SONIC	PLS2
FWSTD2	FWAT-A	M305A ACOUSTIC DITS-2	SONIC	PLS2
FWSTDA	FWAT-A	M305A ACOUSTIC DITS-2	SONIC	PLS2
G-CBL	CBLEB	CEMENT BOND TOOL	CEMENT EVAL.	CHLS
G_GR	GO GR	DDL GAMMA RAY TOOL	NUCLEAR	PLS2
G_SFT4	G_SFT4	G SERIES SFT4	SAMPLING	PLS2

Mnemonic	Tool	Description	Type	Code
G_TEMP	TMP-IC	G_SERIES TEMPERATURE TOOL	PRODUCTION	PLS2
GAMMA	NGRT-A	GAMMA RAY TOOL	NUCLEAR	PLS2
GR	GRDC	NATURAL GAMMA	NUCLEAR	XL1K
GR	GRGA	NATURAL GAMMA—SLIM HOLE	NUCLEAR	XL1K
GR	GRHA	NATURAL GAMMA	NUCLEAR	XL1K
GR	GRIA	NATURAL GAMMA	NUCLEAR	XL1K
GR	GRLA	NATURAL GAMMA	NUCLEAR	XL1K
GR	GRRA	NATURAL GAMMA—MUX PL	NUCLEAR	XL1K
GR	GRRB	NATURAL GAMMA—MUX PL	NUCLEAR	XL1K
GR_DC	COSMOS	DC NATURAL GAMMA	NUCLEAR	PLS2
GR_DC1	BELL	DC NATURAL GAMMA	NUCLEAR	PLS2
GR_DC2	COSMOS	DC NATURAL GAMMA	NUCLEAR	PLS2
GR_DN	M507	MULTICHANNEL GAMMA RAY	NUCLEAR	PLS2
GR_DSN	M507	MULTICHANNEL GAMMA RAY	NUCLEAR	PLS2
GR_GP	GPL	GPL NATURAL GAMMA	NUCLEAR	PLS2
GR_HG	HGNC-A	HGNC NATURAL GAMMA	NUCLEAR	PLS2
GR_MC1	M507	MULTICHANNEL GAMMA RAY	NUCLEAR	PLS2
GR_MC2	M507	MULTICHANNEL GAMMA RAY	NUCLEAR	PLS2
GR_PIB	COSMOS	DC NATURAL GAMMA—PIB	NUCLEAR	PLS2
GR_PL	MC	NATURAL GAMMA	NUCLEAR	PLS2
GR_SG	SGNC-A	SGNC NATURAL GAMMA	NUCLEAR	PLS2
GRAD_P	SONDEX	M/C SONDEX GRADIOMANOMETER	PRODUCTION	PLS2
GRAVEL	GRAVEL	GRAVEL PACK ANALYSIS	ANALYSIS	PLS2
GRPERF	M157	M157 GAMMA PERFORATOR	PRODUCTION	PLS2
HDIL	HDIL-A	HOSTILE DUAL INDUCTION	RESISTIVITY	PLS2
HDIL	HDILAA	HOSTILE DUAL INDUCTION	RESISTIVITY	XL1K
HDSN	HDSN-A	HOSTILE DUAL SPACED NEUTRON	NUCLEAR	PLS2
HDSN	HDSNAA	HOSTILE DUAL SPACED NEUTRON	NUCLEAR	XL1K
HECT	HECT-A	HOSTILE FOUR ARM CALIPER	CALIPER	PLS2
HECT	HECTAA	HOSTILE FOUR ARM CALIPER	CALIPER	XL1K
HEDNAV	HEDT	HEDT NAVIGATION PACKAGE(G)	DIPMETER	PLS2
HEDT	HEDT-A	HOSTILE ENV. DIPMETER TOOL	DIPMETER	PLS2
HETS	HETSAA	HOSTILE TELEMETRY SUB—DITS	TELEMETRY	XL1K
HFDT	HFDT	DITS HIGH FREQUENCY DIELECTRIC	RESISTIVITY	PLS2
HFDT	HFDTAA	DITS HIGH FREQUENCY DIELECTRIC	RESISTIVITY	XL1K
HFDTAN	HFDT	DITS HIGH FREQUENCY DIELECTRIC	RESISTIVITY	PLS2
HFWS	HFWSAA	HOSTILE FULL WAVE SONIC	SONIC	XL1K
HFWS2	HFWS-A	HOSTILE SONIC—LONG SPACE 2TR	SONIC	PLS2
HFWS28	HFWS-A	HOSTILE SONIC—LONG SPACE 2TR	SONIC	PLS2
HFWS2A	HFWS-A	HOSTILE SONIC—LONG SPACE 2TR AMPLITUDE	SONIC	PLS2
HFWS4	HFWS-A	HOSTILE SONIC—SHORT SPACE	SONIC	PLS2
HFWS4A	HFWS4	HEST ACOUSTIC SHORT BOND MODE	SONIC	PLS2
HFWSA	HFWS-A	HOSTILE SONIC—FULL WAVE (A)	SONIC	PLS2
HFWSA8	HFWS-A	HOSTILE SONIC—FULL WAVE (A)	SONIC	PLS2
HGNI	HNGIAA	HOSTILE CCL / GAMMA / NEUTRON INST.	AUXILIARY	XL1K
HMST	HMST-A	HMST FORMATION TESTER	SAMPLING	PLS2
HMSTQ	HMSTQ	HYBRID MST—QUARTZ TRANSDUCER	SAMPLING	PLS2
HNGR	HNGR-A	HOSTILE GAMMA DETECTOR	NUCLEAR	PLS2
HNGR	HNGRAA	HOSTILE GAMMA DETECTOR	NUCLEAR	XL1K

Mnemonic	Tool	Description	Type	Code
HPDC	HPDCAA	HSDL—DENSITY DECENTRALIZER	CENTRALIZER	XL1K
HPDC_D	HPDCD-A	DENSITY DECENTRALIZER	CALIPER	PLS2
HPDC_N	HPDCN-A	NEUTRON DECENTRALIZER	CALIPER	PLS2
HPDL	HPDLAA	HSDL—POWERED DECENTRALIZER—LOWER	NUCLEAR	XL1K
HPDU	HPDUAA	HSDL—POWERED DECENTRALIZER—UPPER	NUCLEAR	XL1K
HRI	HRI	HIGH RESOLUTION INDUCTION	RESISTIVITY	PLS2
HRI	HRIBA	HIGH RESOLUTION INDUCTION	RESISTIVITY	XL1K
HRI	HRICA	HIGH RESOLUTION INDUCTION	RESISTIVITY	XL1K
HRTT	HRTTAA	DITS TEMPERATURE TOOL—INLINE	TEMPERATURE	XL1K
HRTTB	HRTT-A	DITS TEMPERATURE TOOL—BOTTOM	TEMPERATURE	PLS2
HRTTI	HRTT-A	DITS TEMPERATURE TOOL—INLINE	TEMPERATURE	PLS2
HSDI	HSDIAA	HOSTILE SPECTRAL DENSITY INST	AUXILIARY	XL1K
HSDL_I	HSDLI-A	HOSTILE SDL INLINE	NUCLEAR	PLS2
HSDL_M	HSDLM-A	HOSTILE SDL MANDREL	NUCLEAR	PLS2
HSDM	HSDMAA	HOSTILE SPECTRAL DENSITY MANDREL	NUCLEAR	XL1K
HSDP	HSDPAA	HOSTILE SPECTRAL DENSITY PAD	NUCLEAR	XL1K
HSN	HSN-A	HOSTILE SHORT NORMAL	RESISTIVITY	PLS2
HYD	HYDFA	HYDRO TOOL CENTER SAMPLE	PRODUCTION	XL1K
HYD	HYDFB	HYDRO TOOL CENTER SAMPLE	PRODUCTION	XL1K
HYD	HYDFC	HYDRO TOOL CENTER SAMPLE	PRODUCTION	XL1K
IEL	IELAA	INDUCTION ELECTRIC TOOL	RESISTIVITY	XL1K
IEL	IELBA	INDUCTION ELECTRIC TOOL	RESISTIVITY	XL1K
IEL	IELCA	INDUCTION ELECTRIC TOOL	RESISTIVITY	XL1K
IEL	IELDA	INDUCTION ELECTRIC TOOL	RESISTIVITY	XL1K
IEL	IELDB	INDUCTION ELECTRIC TOOL	RESISTIVITY	XL1K
IEL	IELEA	INDUCTION ELECTRIC TOOL	RESISTIVITY	XL1K
IEL	IELFA	INDUCTION ELECTRIC TOOL—(2 3/4 INCH TOOL)	RESISTIVITY	XL1K
IEL	IELGA	INDUCTION ELECTRIC TOOL	RESISTIVITY	XL1K
LFD	LFD	M305B MONO/DIPOLE—FILTER SELECT	SONIC	PLS2
LFD_D2	LFD	M305B LF MONO P TOOL; DIPOLE XMTR—HIGH DATA RATE	SONIC	PLS2
LFD2DT	LFD	M305B LF DIPOLE TOOL; DIPOLE XMTR	SONIC	PLS2
LFD2MT	LFD	M305B LF DIPOLE TOOL; MONO P XMTR	SONIC	PLS2
LFDDT	LFD	M305B LF DIPOLE TOOL; DIPOLE XMTR	SONIC	PLS2
LFDMT	LFD	M305B LF DIPOLE TOOL; MONO P XMTR	SONIC	PLS2
LFDT	LFDT	LOW FREQ DIPOLE ACOUSTIC	SONIC	PLS2
LFDT	LFDTAA	LOW FREQ DIPOLE ACOUSTIC	SONIC	XL1K
LFDT8	LFDT	LOW FREQ DIPOLE ACOUSTIC	SONIC	PLS2
LFDTDT	LFDT	LF DIPOLE TOOL; DIPOLE XMTR.	SONIC	PLS2
LFDTM	LFDT	LFDT ACOUSTIC; MONOPOLE XMTR.	SONIC	PLS2
LFDTMT	LFDT	LF DIPOLE TOOL; MONO P XMTR.	SONIC	PLS2
LFS	LFD	M305B LSS / FWS CONCURRENT 6″ RES @ 34 FPM	SONIC	PLS2
LFS_D2	LFD	M305B LSS / FWS CONCURRENT 3″ RES @ 34 FPM (HIGH DATA RATE)	SONIC	PLS2
LFS_Q2	LFD	M305B LSS / FWS CONCURRENT 6″ RES @ 34 FPM—QUAD STACK	SONIC	PLS2

Mnemonic	Tool	Description	Type	Code
LIDA	LIDA	LITHOLOGY IDENTIFICATION ANALYSIS	ANALYSIS	PLS2
LL3	LL3	LATEROLOG 3	RESISTIVITY	PLS2
LSS	FWAT-A	M305B ALL RCVRS FIXED GAIN—DT ONLY	SONIC	PLS2
LSS	LSSEA	LONG SPACED SONIC—(WVF)	SONIC	XL1K
LSS	LSSIA	LONG SPACED SONIC—(WVF)	SONIC	XL1K
LSS	LSSKA	LONG SPACED SONIC—(WVF)	SONIC	XL1K
LSS	LSSLA	LONG SPACED SONIC—(WVF)	SONIC	XL1K
LSS_D2	FWAT-A	M305B 4 RCVRS FIXED GAIN—QUAD STACK	SONIC	PLS2
LSS_FF	FWAT-A	M305B 3 RCVRS FIXED GAIN—FRAC FINDER	SONIC	PLS2
LSSAFW	M305	M305 Long Spaced Sonic—Mode	SONIC	PLS2
LSSFW	M305	M305 COMBINATION Sonic—Mode	SONIC	PLS2
M/CHD	DUMMY	M/C SERVICE HEADER + ANALOG	DUMMY TOOL	PLS2
M_FLOW	PCU	M/C SONDEX FULLBORE FLOWMETER	PRODUCTION	PLS2
M_GRAD	PCU	M/C SONDEX GRADIOMANOMETER	PRODUCTION	PLS2
M_LOG1	HFDT	HFDT MICROLOG	RESISTIVITY	PLS2
M202	M202	MULTICHANNEL CALIPER	CALIPER	PLS2
M202	M202AA	CALIPER—TWO ARM—DITS BOW SPRING	CALIPER	XL1K
M213	M213	MULTICHANNEL CALIPER	CALIPER	PLS2
M214	M214	M/C CASING COLLAR LOCATOR	PRODUCTION	PLS2
M271	M271	MULTICHANNEL BOND TOOL	CEMENT EVAL.	CHLS
M271_C	M271	MULTICHANNEL SONIC CASED HOLE	CEMENT EVAL.	PLS2
M271_O	M271	MULTICHANNEL SONIC OPEN HOLE	SONIC	PLS2
M271D	M271	M/C M271 DIGITAL	CEMENT EVAL.	PLS2
M271D2	M271	M271 DIGITAL—4 RCVR TO TAPE	CEMENT EVAL.	PLS2
M307	M307	CEMENT BOND TOOL (SINGLE CHAN)	CEMENT EVAL.	CHLS
M307A	M307	SINGLE CHANNEL SONIC AMPLITUDE	CEMENT EVAL.	PLS2
M307D	M307D	DIGITAL SINGLE CHANNEL SONIC	CEMENT EVAL.	PLS2
M307V	M307	SINGLE CHANNEL SONIC VELOCITY	SONIC	PLS2
M310	M310	MULTICHANNEL DUAL INDUCTION	RESISTIVITY	PLS2
M331	M331	M/C DUAL SPACED NEUTRON	NUCLEAR	PLS2
M333	M507	MULTICHANNEL GAMMA RAY	NUCLEAR	PLS2
M334	M334	MULTICHANNEL GAMMA RAY	NUCLEAR	PLS2
M507	M507	MULTICHANNEL GAMMA RAY	NUCLEAR	PLS2
M507M	M507M	MULTICHANNEL GAMMA RAY	NUCLEAR	PLS2
M904C	904-C	MULTI-CHANNEL DUAL GUARD	RESISTIVITY	PLS2
MAC	MAC	MC MULTI-ARM CALIPER TOOL	CALIPER	PLS2
MAC	MACAA	MULTI-ARM CALIPER TOOL	CALIPER	XL1K
MAC	MACAB	MULTI-ARM CALIPER TOOL	CALIPER	XL1K
MAC	MACAC	MULTI-ARM CALIPER TOOL	CALIPER	XL1K
MAC	MACBA	MULTI-ARM CALIPER TOOL	CALIPER	XL1K
MAC	MACBB	MULTI-ARM CALIPER TOOL	CALIPER	XL1K
MAC	MACBC	MULTI-ARM CALIPER TOOL	CALIPER	XL1K
MAC	MACBD	MULTI-ARM CALIPER TOOL	CALIPER	XL1K
MAC	MACBE	MULTI-ARM CALIPER TOOL	CALIPER	XL1K
MAC	MACCA	MULTI-ARM CALIPER TOOL	CALIPER	XL1K
MAC	MACCB	MULTI-ARM CALIPER TOOL	CALIPER	XL1K

Mnemonic	Tool	Description	Type	Code
MAC	MACCC	MULTI-ARM CALIPER TOOL	CALIPER	XL1K
MAC	MACDA	MULTI-ARM CALIPER TOOL	CALIPER	XL1K
MAC	MACDB	MULTI-ARM CALIPER TOOL	CALIPER	XL1K
MAC	MACDC	MULTI-ARM CALIPER TOOL	CALIPER	XL1K
MAC	MACEA	MULTI-ARM CALIPER TOOL	CALIPER	XL1K
MAC	MACEB	MULTI-ARM CALIPER TOOL	CALIPER	XL1K
MAC	MACEC	MULTI-ARM CALIPER TOOL	CALIPER	XL1K
MAC_DC	MAC-AA	MULTI-ARM CALIPER TOOL	CALIPER	PLS2
MACTDC	MAC-AA	MULTI-ARM CALIPER TOOL	CALIPER	PLS2
MC	MC	MULTICHANNEL PPM SUB / CIT / DSN / GEN PURPOSE	COMPLETION	CHLS
MC	MCAA	MULTICHANNEL PPM SUB (DUMMY TOOL)	DUMMY TOOL	XL1K
MC_CCL	M214	M/C CASING COLLAR LOCATOR	COMPLETION	PLS2
MC_DSN	M265A	M/C DUAL SPACED NEUTRON	NUCLEAR	PLS2
MC_GN	DUMMY	DUMMY MC	DUMMY TOOL	PLS2
MC_PL	DUMMY	PRODUCTION MULTICHANNEL SUB	TELEMETRY	PLS2
MCCEB	DUMMY	MC CABLE ELECTRODE BRIDLE	AUXILIARY	PLS2
MCCL	M214	M/C CASING COLLAR LOCATOR	COMPLETION	PLS2
MCDSN	M265A	M/C DUAL SPACED NEUTRON M265A	NUCLEAR	PLS2
MCFRXO	MGRD	MULTI_CHANNEL MICRO GUARD	RESISTIVITY	PLS2
MCGRD	M320	MICRO GUARD-DLLT-A	RESISTIVITY	PLS2
MDIP	M243A	MULTICHANNEL DIPMETER MANDREL	DIPMETER	PLS2
MEL	MELCA	MICRO ELECTRIC LOG—DC	RESISTIVITY	XL1K
MEL	MELDA	MICRO ELECTRIC LOG—DC	RESISTIVITY	XL1K
MEL	MELDB	MICRO ELECTRIC LOG—DC	RESISTIVITY	XL1K
MEL	MELDC	MICRO ELECTRIC LOG—DC	RESISTIVITY	XL1K
MEL_DC	CA	MICRO ELECTRIC LOG—DC	RESISTIVITY	PLS2
MELCAL	MEL	DDL MEL CALIPER	CALIPER	PLS2
MELPUL	DA,DB,DC	MICRO ELECTRIC LOG—DDL—PULSE	RESISTIVITY	PLS2
MGCAL	MGRD	MULTI-CHANNEL MICRO GUARD CAL	RESISTIVITY	PLS2
MGRD	MGRDAA	MICRO GUARD RESISTIVITY-DLLT-A	RESISTIVITY	XL1K
MICLGC	MSFL	MICROLOG ON DITS MSFL	RESISTIVITY	PLS2
MICLOG	SDLT-A	MICROLOG—SDLT	RESISTIVITY	PLS2
MLL	MLLAA	MICROLATEROLOG RESISTIVITY	RESISTIVITY	XL1K
MLL	MLLBA	MICROLATEROLOG RESISTIVITY	RESISTIVITY	XL1K
MLL	MLLCA	MICROLATEROLOG RESISTIVITY	RESISTIVITY	XL1K
MSFCAL	MSFL	MSFL CALIPER	CALIPER	PLS2
MSFL	MSFL	DDL MICRO-SPHERICALLY FOCUSED	RESISTIVITY	PLS2
MSFL	MSFLCA	MICRO-SPHERICALLY FOCUSED-DDL	RESISTIVITY	XL1K
MSFL	MSFLDA	MICRO-SPHERICALLY FOCUSED-DDL	RESISTIVITY	XL1K
MSFL	MSFLDB	MICRO-SPHERICALLY FOCUSED-DDL	RESISTIVITY	XL1K
MSFL	MSFLDC	MICRO-SPHERICALLY FOCUSED-DDL	RESISTIVITY	XL1K
MSFL	MSFLEA	MICRO-SPHERICALLY FOCUSED-DDL	RESISTIVITY	XL1K
MSFL	MSFLFA	MICRO-SPHERICALLY FOCUSED-DDL	RESISTIVITY	XL1K
MSFL40	MSFL40	DDL MSFL LOW RES OPTION X40	RESISTIVITY	PLS2
MSFLM	M320	MSFL MANDREL WITH DITS DLL	RESISTIVITY	PLS2
MSFT	MSFTAA	MICRO-SPHERICALLY FOCUSED-DITS	RESISTIVITY	XL1K
MSFT	MSFTLM	MICRO-SPHERICALLY FOCUSED-DITS	RESISTIVITY	XL1K
MSFT	MSFTUM	MICRO-SPHERICALLY FOCUSED-DITS	RESISTIVITY	XL1K

Mnemonic	Tool	Description	Type	Code
MTRACM	MC-CSNG	MC CSNG MINI TOOL	NUCLEAR	PLS2
MUX	MUXLA	MUX TELE FOR NEUTRON—SUPER STACK	TELEMETRY	XL1K
MUXPL	MUXPL	MUXB-PL DATA TOOL	PRODUCTION	CHLS
NAVS	NAVS-A	FACT NAVIGATION SUB	AUXILIARY	PLS2
NE_HG	HGNC-A	NEUTRON HE3 DETECTOR	NUCLEAR	PLS2
NE_SG	SGNC-A	NEUTRON HE3 DETECTOR	NUCLEAR	PLS2
NEU	NEUAA	CASE HOLE NEUTRON (1 IN.)—SINGLE DETECTOR	NUCLEAR	XL1K
NEU	NEUBA	CASE HOLE NEUTRON (1.68 IN.)—SINGLE DETECTOR	NUCLEAR	XL1K
NEU	NEUCA	CASE HOLE NEUTRON (3.5 IN.)—SINGLE DETECTOR	NUCLEAR	XL1K
NEUDC	COSMOS	DC NEUTRON (15″ SOURCE SUB)	AUXILIARY	PLS2
NEUDC1	COSMOS	DC NEUTRON (15″ SOURCE SUB)	AUXILIARY	PLS2
NEUPIB	COSMOS	DC NEUTRON (15″ SOURCE SUB)—(PIB)	AUXILIARY	PLS2
NGRT	NGRTAA	GAMMA RAY TOOL-DITS	NUCLEAR	XL1K
PCK	PCK	PANEL—XL1-GRAVEL PACK	DUMMY TOOL	XL1K
PCUA	PCUA	MULTICHANNEL PRODUCTION LOGGING TOOL	PRODUCTION	CHLS
PCUB	PCUB	MULTICHANNEL PRODUCTION LOGGING TOOL	PRODUCTION	CHLS
PET	PET	PULSE ECHO TOOL (MUX-B)	CEMENT EVAL.	CHLS
PET	PET	PULSE ECHO TOOL (MUX-B)	CEMENT EVAL.	PLS2
PET	PETAA	PULSE ECHO TOOL (MUX-B)	CEMENT EVAL.	XL1K
PET	PETBB	PULSE ECHO TOOL (MUX-B)	CEMENT EVAL.	XL1K
PET/CB	PET/CB	PULSE ECHO TOOL (MUX-B)	CEMENT EVAL.	CHLS
PET_C	PET_C	PULSE ECHO TOOL (MUX-B)—(QC JOINTS)	CEMENT EVAL.	PLS2
PETGR	UGR	PET TOOL GAMMA RAY	NUCLEAR	PLS2
PGPP		PANEL—GENERAL PURPOSE DDL HEADER	DUMMY TOOL	XL1K
PIT	PIT	DITS PIPE INSPECTION TOOL	CASING INSPECT.	PLS2
PIT8	PIT	DITS PIPE INSPECTION TOOL-8 PAD	CASING INSPECT.	PLS2
PL_HD	MC	PROD. LOGGING HEADER + ANALOG	DUMMY TOOL	PLS2
PLA	PLA	PRODUCTION LOG ANALYSIS	ANALYSIS	PLS2
PQ_PL	PCU	PRESSURE-(PETRO-QUARTZ)	COMPLETION	PLS2
PR_PL	PCU	PRESSURE—(WELL TEST)	COMPLETION	PLS2
PSGT	PSGT-A	PULSE SPECTRAL GAMMA TOOL	NUCLEAR	PLS2
PSGT	PSGTAA	PULSE SPECTRAL GAMMA TOOL	NUCLEAR	XL1K
PSGTMD	ANAL	PSGT TMD PROCESSING	ANALYSIS	PLS2
PSYS		PANEL—DDL SYSTEM GENERAL STATUS	DUMMY TOOL	XL1K
QPG	QPGAA	QUARTZ PRESSURE GUAGE	PRODUCTION	XL1K
QPG	QPGBA	QUARTZ PRESSURE GUAGE	PRODUCTION	XL1K
ROTASC	ROTASC	ROTASCAN (MUST BE RUN WITH TSCAN 1 11/16)	NUCLEAR	CHLS
SASHA	SASHA	SANDY SHALE ANALYSIS MODEL	ANALYSIS	PLS2
SDDT	NAV-DDT	STANDALONE DITS DIRECTIONAL	AUXILIARY	PLS2

Mnemonic	Tool	Description	Type	Code
SDDT	SDDTA	STANDALONE DITS DIRECTIONAL	AUXILIARY	XL1K
SDL	SDLT-A	SPECTRAL DENSITY LOG	NUCLEAR	PLS2
SDL8	SDLT-A B	SPECTRAL DENSITY LOG (8 BIT)	NUCLEAR	PLS2
SDLT	SDLTAA	SPECTRAL DENSITY LOG—(8 BIT)	NUCLEAR	XL1K
SDLT	SDLTBA	SPECTRAL DENSITY LOG—(12 BIT)	NUCLEAR	XL1K
SDLT	SDLTBB	SPECTRAL DENSITY—(FLOATING BODY)	NUCLEAR	XL1K
SED	SED-C	SIX ELECTRODE DIPMETER—DITS	DIPMETER	PLS2
SED	SEDAO	SIX ELECTRODE DIPMETER—(OIL BASE)	DIPMETER	XL1K
SED	SEDAW	SIX ELECTRODE DIPMETER—(WATER BASE)	DIPMETER	XL1K
SED	SEDBO	SIX ELECTRODE DIPMETER—(OIL BASE) DITS	DIPMETER	XL1K
SED	SEDBW	SIX ELECTRODE DIPMETER—(WATER BASE) DITS	DIPMETER	XL1K
SED	SEDS	SIX ELECTRODE DIPMETER—(SONIC COMBO) DITS	DIPMETER	XL1K
SEDNAV	SED-C	SED NAVIGATION PACKAGE (G)	AUXILIARY	PLS2
SEDRES	SED-RES	SED HI-RESOLUTION RESISTIVITY	DIPMETER	PLS2
SFT	SFTDA	SEQ. FORMATION TESTER—DDL—TYPE III	SAMPLING	XL1K
SFT	SFTDB	SEQ. FORMATION TESTER—DDL—TYPE III	SAMPLING	XL1K
SFT	SFTDC	SEQ. FORMATION TESTER—DDL—TYPE III	SAMPLING	XL1K
SFT	SFTEA	SEQ. FORMATION TESTER—DDL—TYPE III	SAMPLING	XL1K
SFT	SFTEB	SEQ. FORMATION TESTER—DDL—TYPE III	SAMPLING	XL1K
SFT	SFTFA	SEQ. FORMATION TESTER—DDL—SLIM HOLE	SAMPLING	XL1K
SFT	SFTGA	SEQ. FORMATION TESTER—DDL—TYPE IV	SAMPLING	XL1K
SFT	SFTGB	SEQ. FORMATION TESTER—DDL—TYPE IV	SAMPLING	XL1K
SFT	SFTHA	SEQ. FORMATION TESTER—DDL—TYPE IV	SAMPLING	XL1K
SFT	SFTJA	SEQ. FORMATION TESTER—DDL—TYPE IV	SAMPLING	XL1K
SFT	SFTLA	SEQ. FORMATION TESTER—DDL—TYPE IV	SAMPLING	XL1K
SFT	SFTT-A	SEQUENTIAL FORMATION TESTER	SAMPLING	PLS2
SFT4	SFT4	SFT4 PETRO QUARTZ	SAMPLING	PLS2
SFTC	SFTC	SFT PETRO QUARTZ TOOL (SFTT-C)	SAMPLING	PLS2
SFTI	SFTIAA	SEQ. FORMATION TESTER—DITS INSTRUMENT	SAMPLING	XL1K
SFTPQ	SFTT-B	SFT PETRO QUARTZ TOOL	SAMPLING	PLS2
SFTT	SFTTAA	SEQ. FORMATION TESTER—DITS	SAMPLING	XL1K
SGR	SGR	SPECTRAL GAMMA RAY (DIO #9)	NUCLEAR	PLS2
SGR	SGRAC	SPECTRAL GAMMA RAY TOOL	NUCLEAR	XL1K
SGR	SGRBA	SPECTRAL GAMMA RAY TOOL	NUCLEAR	XL1K

Mnemonic	Tool	Description	Type	Code
SGRD	DSGT-A B	SHORT GUARD RESISTIVITY SUB—(DILT)	RESISTIVITY	PLS2
SGRD	SGRDAA	SHORT GUARD RESISTIVITY SUB—(DILT)	RESISTIVITY	XL1K
SHIVA	SHIVA	SIX ELECTRODE DIPMETER ANALYSIS	ANALYSIS	PLS2
SHIVA4	SHIVA4	FOUR ELECTRODE DIPMETER ANALYSIS	ANALYSIS	PLS2
SHVOMN	SHVOMN	SIX ELECTRODE DIPMETER OMNIPLOT ANALYSIS	ANALYSIS	PLS2
SILT	SILT	SLIM LINE INDUCTION (2 3/4″)	RESISTIVITY	PLS2
SLD	SLD-A	SPECTRAL LITHO-DENSITY—(DIO #7)	NUCLEAR	PLS2
SLD	SLDBA	SPECTRAL LITHO-DENSITY TOOL	NUCLEAR	XL1K
SLD	SLDBB	SPECTRAL LITHO-DENSITY TOOL	NUCLEAR	XL1K
SLD	SLDDA	SPECTRAL LITHO-DENSITY TOOL— (MUX-B)	NUCLEAR	XL1K
SLD_BB	SLD-A	SPECTRAL LITHO-DENSITY—(DIO #7)	NUCLEAR	PLS2
SLDCAL	SLDCAL	SLD CALIPER	CALIPER	PLS2
SNP	SNP	SIDEWALL NEUTRON	NUCLEAR	PLS2
SNP	SNPAA	SIDEWALL NEUTRON	NUCLEAR	XL1K
SNP	SNPBA	SIDEWALL NEUTRON	NUCLEAR	XL1K
SNP	SNPBB	SIDEWALL NEUTRON	NUCLEAR	XL1K
SNPCAL	SNPCAL	SNP CALIPER	NUCLEAR	PLS2
SOTX	CALIPER	DSTU STANDOFF TOOL—(CALIPER)	AUXILIARY	PLS2
SP_HRI	HRI	ANALOG SP—MODIFIED HRI TOOL	AUXILIARY	PLS2
SPC	SPCAA	PRODUCTION LOGGING CENTRALIZER	CENTRALIZER	XL1K
SPIN	BELL	DC SPINNER	PRODUCTION	PLS2
SPL	SPLGB	PRODUCTION LOGGING MUX-B SUB	PRODUCTION	XL1K
SPN_PL	PL	SPINNER	PRODUCTION	PLS2
SPT	SPTCA	PAINE PRESSURE TOOL	PRODUCTION	XL1K
SPT	SPTCB	PAINE PRESSURE TOOL	PRODUCTION	XL1K
SPT	SPTCC	PAINE PRESSURE TOOL	PRODUCTION	XL1K
SSNT	SILT	SLIM LINE SHORT NORM (2 3/4″)	RESISTIVITY	PLS2
SSS	FWAT-A	M305B 2 RCVR FIXED GAIN SHORT SPACE TIP	SONIC	PLS2
STOP	STOP	STOP CHECKS—PL PANEL	PRODUCTION	XL1K
SWN	M166	SIDEWALL NEUTRON	NUCLEAR	PLS2
TAC	TACBA	TWO ARM CALIPER—DDL	CALIPER	XL1K
TAC	TACCA	TWO ARM CALIPER—DDL	CALIPER	XL1K
TEMP	BELL	DC TEMPERATURE	TEMPERATURE	PLS2
TEMPSW	SWN	SWN TEMPERATURE	TEMPERATURE	PLS2
TIP	TIPAA	DDL SUPER STACK TOP ISOLATION SUB	AUXILIARY	XL1K
TMD	M395	THERMAL MULTI-GATE DECAY (TMD)	NUCLEAR	CHLS
TMD	M395	THERMAL MULTI-GATE DECAY (TMD)	NUCLEAR	PLS2
TMD	TMDWX	THERMAL MULTI-GATE DECAY (TMD)	NUCLEAR	XL1K
TMDGR	M395	TMD GAMMA RAY	NUCLEAR	PLS2
TMDL	TMDL	THERMAL MULTI-GATE DECAY LITHOLOGY TOOL	NUCLEAR	CHLS
TMDL	TMDLAA	THERMAL MULTI-GATE DECAY LITHOLOGY TOOL	NUCLEAR	XL1K
TMP	TMP1A	TEMPERATURE PANEL—GENERAL	COMPLETION	XL1K
TMP	TMP1B	TEMPERATURE PANEL—GENERAL	COMPLETION	XL1K
TMP	TMP1C	TEMPERATURE PANEL—GENERAL	COMPLETION	XL1K
TMP_PL	PCU	TEMPERATURE—(MC)	TEMPERATURE	PLS2

Mnemonic	Tool	Description	Type	Code
TPH_PL	PCU	TEMPERATURE HI RES	TEMPERATURE	PLS2
TPL	TPLAA	TOOL PUSHER PANEL—MUX A/B	AUXILIARY	XL1K
TPL	TPLBB	TOOL PUSHER PANEL—MUX A/B	AUXILIARY	XL1K
TRACER	CSNGG-A	DITS TRACER SCAN (GRAPHITE)	NUCLEAR	PLS2
TRACER	TRACER	MC TRACERSCAN—MINITOOL	NUCLEAR	CHLS
TRACMC	MC-CSNG	MC CSNG LARGE TOOL—GRAPHITE	NUCLEAR	PLS2
TTRACE	CSNGT-A	DITS TRACER SCAN (TITANIUM)	NUCLEAR	PLS2
TTRACM	MC-CSNG	MC CSNG LARGE TOOL—TITANIUM	NUCLEAR	PLS2
TTTC	TTTCAA	THRU TUBING TELEMETRY CARTRIDGE	TELEMETRY	XL1K
TVS	TVSAA	CIRCUMFERENCIAL ACOUSTIC SCANNING TOOL	IMAGING	XL1K
TVS	TVSBA	CIRCUMFERENCIAL ACOUSTIC SCANNING TOOL	IMAGING	XL1K
V_REG	V_REG	VOLTAGE REGULATOR SUB	AUXILIARY	PLS2
VCR	VCR	VCR—TOOL SIMULATOR MODE	DUMMY TOOL	XL1K
VRS	VRSAA	VOLTAGE REGULATOR SUB	AUXILIARY	XL1K
VRS	VRSCA	VOLTAGE REGULATOR SUB	AUXILIARY	XL1K
WVF	WVFAA	WAVEFORM PANEL-(LSS)—DDL GENERAL	AUXILIARY	XL1K
XYC	XYCAA	X-Y CALIPER TOOL	CALIPER	XL1K
XYC	XYCBA	X-Y CALIPER TOOL	CALIPER	XL1K
XYC	XYCCA	X-Y CALIPER TOOL	CALIPER	XL1K
XYC	XYCDA	X-Y CALIPER TOOL	CALIPER	XL1K
XYC	XYCDB	X-Y CALIPER TOOL	CALIPER	XL1K
XYC	XYCXX	X-Y CALIPER TOOL—(CALIF. OPTION)	CALIPER	XL1K

Baker Atlas Tool Mnemonics

Tool	Description
3VSP	3-AXIS GEOPHONE
4CAL	4 ARM CALIPER
AC	ACOUSTIC LOG
AGN	AIRGUN
AP	AUTOMATIC DIFFERENTIAL PRESSURE VALVE
BAL	BOND ATTENUATION LOG
BPS	FMT BYPASS SUB
BRDL	CABLEHEAD WITH 85 FEET OF BRIDLE MATERIAL AND SP ELECTRODE
CAL	CALIPER
CBIL	CIRCUMFERENTIAL BOREHOLE IMAGING LOG
CBL	CEMENT BOND LOG
CCL	CASING COLLAR LOCATOR
CDB	DUMP BAILER
CDL	COMPENSATED DENSITY LOG
CENT	CENTRALIZER
CH	CABLEHEAD WITH SP ELECTRODE BUTTON
CHL	CHLORINE LOGGING TOOL
CHTS	CABLEHEAD TENSION MEASUREMENT
CMI	COMPACTION MONITORING INSTRUMENT
CN	COMPENSATED NEUTRON LOG
CO	CARBON/OXYGEN LOG
DAC	DIGITAL ARRAY ACOUSTILOG LOG
DAL	DIGITAL ACOUSTIC LOG
DCEN	DE-CENTRALIZER
DEL2	200 MHz DIELECTRIC LOG
DEL4	47 MHz DIELECTRIC LOG
DEN	NON-COMPENSATED DENSITY LOG
DFS	DISPLACEMENT FLUID SAMPLER
DHPA	DOWNHOLE POWER ADAPTOR
DIFL	DUAL INDUCTION FOCUS LOG
DIP	DIPLOG
DLL	DUAL LATEROLOG
DMAG	DIGITAL MAGNELOG
DPIL	DUAL PHASE INDUCTION LOG
DSL	DIGITAL SPECTRALOG
DVRT	DIGITAL VERTILOG
FCON	WELLBORE FLUID CONDUCTIVITY (SALINITY) LOG
FDDP	FLUID DENSITY FROM DIFFERENTIAL PRESSURE LOG
FDN	FLUID DENSITY LOG
FMCS	FLOWMETER—CONTINUOUS SPINNER
FMFI	FLOWMETER—FOLDING IMPELLER
FMT	FORMATION MULTI-TESTER

Tool	Description
FPI	FREE POINT INDICATOR (PIPE RECOVERY)
FS	BOREHOLE FLUID SAMPLER
GP	GRAVEL PACK (SHAKER)
GR	GAMMA RAY LOG
GRC	PRESSURE (GRC STRAIN GAUGE)
GRN	GAMMA RAY/NEUTRON COMBINATION
HDIL	HIGH DEFINITION INDUCTION LOG
HDIP	6-ARM (HEX) DIPLOG
HDLL	HIGH DEFINITION LATEROLOG
HP	PRESSURE (HP QUARTZ GAUGE)
HOIS	SURFACE SYSTEM PANEL (HOIST/DEPTH)
HTD	HIGH TEMPERATURE DENSITY LOG
HYDL	HYDROLOG
IEL	INDUCTION ELECTROLOG
ISSB	MASS-ISOLATION JOINT
JBSK	JUNK BASKET
KNJT	KNUCKLE JOINT
LL3	LATERALOG (3 ELECTRODE)
M5M7	CROSSOVER POINT FOR M5 TO M7 OR M7 TO M5
MAC	MULTIPLE ARRAY ACOUSTIC LOG
MAC2	MULTIPLE ARRAY ACOUSTIC LOG
MAG	MAGNELOG (ANALOG)
MCFM	MULTI-CAPACITANCE FLOW METER
MFC	MULTI-FINGER CALIPER
MFP	CONTINUOUS MAGNETIC FREE POINT SENSOR
ML	MINILOG
MLL	MICRLATEROLOG
MRIL	MAGNETIC RESONANCE IMAGING LOG
MSI	MULTIPARAMETER SPECROSCOPY INSTRUMENT
MSL	MICRO SPHERICAL LATEROLOG WITH CALIPER
MST	HEAVY-WALLED DRILL PIPE SEVERING WITH SYNCHRONIZED DETONATION
NEU	SINGLE DETECTOR NEUTRON
NFL	TRACER/FLO-LOG
NIR	NEAR INFARED INSTRUMENT
NLL	NEUTRON LIFETIME LOG
NO	NUCLEAR ORIENTATOR
ORIT	ORIENTATION LOG
PDK	PULSE/DECAY PULSED NEUTRON LOG
PFC	PERFORATING FORMATION CORRELATION LOG
PHT	PHOTON LOG
PNHI	PULSED NEUTRON HOLDUP INDICATOR
POS	POWERED ORIENTAION SWIVEL
PROX	PROXIMITY LOG
PRSM	PRISM LOG (MULTIPLE ISOTOPE TRACING)

Tool	Description
RCI	RESERVOIR CHARACTERIZATION INSTRUMENT
RCOR	ROTARY SIDEWALL CORING TOOL
ROLR	ROLLER BODY
RPL	PANEX PRESSURE GAUGE
RPM	MULTI-PURPOSE SMALL DIAMETER PULSED NEUTRON LOGGING INSTRUMENT
SBT	SEGMENTED BOND TOOL
SG	STRAIN GAUGE
SGEO	SURFACE GEOPHONE
SJ	SILVER JET PERFORATOR
SK	SELECTKONE PERFORATOR
SL	SPECTRALOG
SLAP	SIMULTANEOUS LOGGING AND PERFORATING
SLKP	SLIMKONE PERFORATOR
SNKB	SINKER BAR
SON	SONAN (NOISE LOG)
SPCR	SPACER BAR
SPDK	SPECTRAL PULSE/DECAY PULSED NEUTRON LOG
SPSB	SP (SPONTANEOUS POTENTIAL) SUB
SRPL	SURFACE RECORDED PRESSURE LOG
SSP	SEMI-SELECT PERFORATOR
STAR	SIMULTANEOUS ACOUSTIC-RESISTIVITY IMAGING LOG
STFP	PIPE FREE POINT INDICATOR
SUB	GYRO DATA INTERCONNECT
SUB	AC/DC SWITCHING CIRCUIT TO RUN PHOTON & GRAVEL PACK IN COMBO
SWC	SIDEWALL CORGUN
SWN	SIDEWALL NEUTRON
SWVL	SWIVEL
TBFS	THROUGH TUBING FLUID SAMPLER
TBRT	THIN-BED RESISTIVITY LOG
TCAL	THROUGH TUBING CALIPER
TCP	TUBING-CONVEYED PERFORATING
TCR	THROUGH CASING RESISTIVITY
TEMP	TEMPERATURE LOG
TILT	TRANSVERSE INDUCTION LOG
TMFP	MAGNA-TECTOR FREE POINT INDICATOR
TPFM	THERMAL PULSE FLOWMETER
TTRM	TEMPERATURE, TENSION AND MUD RESISTIVITY SUB
VRT	VERTILOG
VSP	VERTICAL SEISMIC PROFILE
VTLN	VERTILINE
WCAL	BOWSPRING 6-ARM CALIPER
WESB	SIDE ENTRY SUB FOR PIPE CONVEYED LOGGING
WHI	WATER HOLD-UP INDICATOR

Tool	Description
WTS	WIRELINE TRANSMISSION SYSTEM DOWNHOLE TELEMETRY REPEATER
WTSP	WIRELINE TRANSMISSION SYSTEM SURFACE PANEL
WTSS	WIRELINE TRANSMISSION SYSTEM SWITCHING SUB
XMAC	CROSS MULTIPOLE ARRAY ACOUSTILOG
ZDL	Z-DENSITY LOG

BIBLIOGRAPHY

Basic Evaluation Techniques

Archie, G. E. "The Electrical Resistivity Log as an Aid in Determining Some Reservoir Characteristics." *Petroleum Technology* 5 (January 1942).

Juhasz, I. "Porosity Systems and Petrophysical Models Used in Formation Evaluation." SPWLA London Chapter Porosity Seminar, 26 April 1988.

Core Analysis

Anderson, W. G. "Wettability Literature Survey Part 1: Rock/Oil/Brine Interactions and the Effects of Core Handling on Wettability." *Journal of Petroleum Technology* 38 (October 1986): 1125–1144.

Rathmell, J. J., Tibbitts, G. A., Gremley, R. B., Warner, H. R., & White, E. K. "Development of a Method for Partially Uninvaded Coring in High Permeability Sandstones." Paper 20413 presented at SPE Annual Technical Conference, Formation Evaluation, New Orleans, June 1990.

Rathmell, J. J., Gremley, R. B., & Tibbitts, G. A. "Field Applications of Low Invasion Coring." Paper 27045 presented at SPE Annual Technical Conference, Houston, 3–5 October 1993.

Skopec, R. A. "Proper Coring and Wellsite Core Handling Procedures: The First Step Towards Reliable Core Analysis." *Journal of Petroleum Technology* 46 (April 1994).

Fluid Replacement Modeling

Gassmann, F. "Uber die elastizitat poroser medien [Elasticity of Porous Media]." *Vierteljahrsschrift der Naturforschenden Gesselschaft* 96, 1951: 1–23.

Logging Tool Types

Wahl, J. S., Tittman, J., & Johnstone, C. W. "The Dual Spacing Formation Density Log." *Journal of Petroleum Technology*, December 1964.

Homing In

de Lange, J. I., & Darling, T. J. *Improved Detectability of Blowing Wells*. Paper 17255 presented at SPE/IADC, March 1988.

Jones, D. L., Hoehn, G. L., & Kuckes, A. F. *Improved Magnetic Model for Determination of Range and Direction of a Blow-Out Well*. SPE paper 14388, September 1985.

Kuckes, A., et al. *An Electromagnetic Survey Method for Directionally Drilling a Relief Well into a Blown-Out Oil or Gas Well*. SPE paper 10946, March 1981.

Mitchell, F. R., Robinson, J. D., et al. "Using Resistivity Measurements to Determine Distance Between Wells." *Journal of Petroleum Technology*, June 1972: 723.

Multimineral/Statistical Models

Mayer, C., & Sibbert, A. "GLOBAL, A New Approach to Computer Processed Log Interpretation." Paper 9341 presented at SPE Annual Tech Conference, 1980.

NMR Logging

Coates, G. R., Gardner, J. S., & Miller, D. L. "Applying Pulsed Echo NMR to Shaly Sand Formation Evaluation." Paper presented at SPWLA Convention, Oklahoma, June 1994.

Slijkerman, W. F., & Hofman, J. P. "Determination of Surface Relaxivity from NMR Diffusion Measurements." *Magnetic Resonance Imaging* 16 (June–July 1998): 541–544.

Production Geology Issues

Shell Internationale Petroleum Maatschappij BV (SIPM) Training Division. Course notes from Shell "Phase 2" Production Geology course, 1986.

Reservoir Engineering Issues

Dake, L. P. *Fundamentals of Reservoir Engineering*. Amsterdam: Elsevier, 1978.

Katz, D. L., et al. *Handbook of Natural Gas Engineering*. New York: McGraw-Hill, 1959.

Marsal, D., for SIPM. Course notes from Shell "Phase 1" Reservoir Engineering course, 1982.

Rock Mechanics Issues

Biot, M. "Theory of Elasticity and Consolidation for a Porous Anisotropic Solid." *Journal of Applied Physics* 26 (1955): 182–185.

Brown, R. J. S., & Kerringa, J. "On the Dependence of the Elastic Properties of a Porous Rock on the Compressibility of the Pore Fluid." *Geophysics* 40 (1975): 608–616.

Saturation/Height Functions

Leverett, M. C. "Capillary Behaviour in Porous Solids." *Transactions of the American Institute of Mining and Metallurgical Engineers* 142 (1941): 152–169.

Shaly Sand Analysis

Clavier, C., Coates, G., & Dumanoir, J. L. "The Theoretical and Experimental Basis for the 'Dual Water' Model for the Interpretation of Shaly Sands." *Society of Petroleum Engineers Journal*, April 1984: 153–167.

Juhasz, I. *The Central Role of Qv and Formation Water Salinity in the Evaluation of Shaly Formations*. Paper presented at the Society of Professional Well Log Analysts 19th Annual Symposium, June 1979.

——. "Normalised Qv: The Key to Shaly Sand Evaluation Using the Waxman-Smits Equation in the Absence of Core Data." Paper presented at SPWLA Symposium, 1981.

Waxman, M. H., & Smits, L. J. M. "Electrical Conductivities in Oil-Bearing Shaly Sands." *Society of Petroleum Engineers Journal*, June 1968: 107–119. Trans AIME 243.

Waxman, M. H., & Thomas, E. C. "Electrical Conductivities in Oil-Bearing Shaly Sands: I. The Relation Between Hydrocarbon Saturation and Resistivity Index. II. The Temperature Coefficient of Electrical Conductivity." *Society of Petroleum Engineers Journal* 14 (1974): 213–225.

Worthington, P. F. "The Evolution of Shaly Sand Concepts in Reservoir Evaluation." *The Log Analyst*, January–February 1985: 23.

Thermal Decay Neutron Interpretation

Clavier, C., Hoyle, W. R., & Meunier, D. "Quantitative Interpretation of TDT Logs, Parts I and II." *Journal of Petroleum Technology*, June 1971.

Thin Beds

Moran, J. H., & Gianzero, S. "Effects of Formation Anisotropy on Resistivity Logging Measurements." *Geophysics* 44 (1978): 1266–1286.

Thomas, E. C., & Steiber, S. J. "The Distribution of Shales in Sandstones and Its Effect Upon Porosity." *Transactions of the 16th Annual SPWLA Logging Symposium*, 1975, paper T.

Well Deviation and Geosteering

Wolff, C. J. M., & de Wardt, J. P. "Borehole Position Uncertainty: Analysis of Measuring Methods and Derivation of Systematic Error Model." *Journal of Petroleum Technology*, December 1981: 2339–2350.

ABOUT THE AUTHOR

Toby Darling was educated at The Pilgrim's School and Winchester College before gaining a scholarship to Wadham College, Oxford, where he graduated with a degree in physics in 1982. He immediately joined Shell International Petroleum Maatschappij (SIPM). After initial training in The Hague, he was posted to Thai Shell from 1983 to 1984, working as a wellsite petroleum engineer. In 1984 he was posted to SIPM in The Hague, working on secondment to the Exploration seismic group; in 1986 he transferred to the Shell Research Laboratories (KSEPL), where he worked on a number of novel logging techniques, including the development of LWD tools. In 1989 he was seconded to Petronas Carigali Sdn Bhd in Miri, Malaysia, where he worked as an operational petrophysicist in the Baram Delta region. In 1992, he moved to Shell Expro in Aberdeen, working as a petrophysicist and team leader for a number of North Sea fields. In 1997 he transferred to Shell Expro's exploration department in London, where he worked until 1999 before leaving Shell to form his own consultancy. Since 1999 he has worked as a consultant to a number of oil companies, most recently on contract to BAPETCO in Cairo. He is married with two sons.

ACKNOWLEDGMENTS

I have been fortunate during most of my career to have worked for a major oil company (Shell) which has been at the forefront of petrophysics. I am therefore grateful to Shell for providing the training and environment that has made this book possible.

I would like to thank BAPETCO and Shell Egypt for permission to use some of their log data in the exercises and for field examples of logs. In particular I would like to thank Erik Jan Uleman for his help in reviewing the inital draft of the book and Mahmoud M. Anwar for his suggestions concerning horizontal well permeability.

I would like to thank Schlumberger, Halliburton, and Baker Atlas for permission to include lists of their tool mnemonics.

INDEX

Printed and bound by CPI Group (UK) Ltd, Croydon, CR0 4YY

03/10/2024

01040433-0005